EARLY SOUTHWEST ORNITHOLOGISTS

Dan L. Fischer

EARLY SOUTHWEST ORNITHOLOGISTS

1528–1900

THE UNIVERSITY OF ARIZONA PRESS   TUCSON

The University of Arizona Press
© 2001 The Arizona Board of Regents
First printing
All rights reserved
♾ This book is printed on acid-free, archival-quality paper.
Manufactured in the United States of America

06   05   04   03   02   01   6   5   4   3   2   1

Library of Congress Cataloging-in-Publication Data

Fischer, Dan L. (Dan Lewis), 1932–
Early southwest ornithologists, 1528–1900, Dan L. Fischer.
    p. cm. – (The Southwest Center Series)
Includes bibliographical references (p.) and index.
ISBN 0-8165-2149-2 (cloth: alk. paper)
    1. Ornithology—Southwest, New—History.
    2. Ornithologists—Southwest, New—History.
    I. Title. II. Series.
QL672.73.S85 F57   2001
598´.0979–dc21   2001001663

British Library Cataloguing-in-Publication Data

A catalogue record for this book is available from the British Library.

DEDICATED TO ALL NATURALISTS—
PAST, PRESENT, AND FUTURE

CONTENTS

FIGURES

Maps

*P*REFACE

 M**Y ACQUAINTANCE WITH** the Southwest began at the age of five when, in 1937, my father brought my ailing mother, my older sister JoAnne, and me to the warm, dry climate of Yuma, Arizona. Occasionally, he would take us just across the Colorado River, to the California side, to play in the great "sand hills" or Imperial Sand Dunes. Once, I remember holding my father's hand as we looked up at the tiny hole of daylight in one of the confinement cells of the dreaded Yuma Territorial Prison (1876–1909).

In 1947, after several more relocations, my interest in birds began to develop. Then, two years later, I was privileged to take a few weekend camping trips to other areas of the southeastern California desert with Edmund C. Jaeger (1887–1983), a wonderful teacher, author, and desert naturalist.

Knowing my interest in birds, Jaeger took me on a trip to the Chuckwalla Mountains, where he had discovered a hibernating Common Poorwill. During the prior three years, Jaeger (1948, 1949) had documented this remarkable event. But this visit proved to be the first time the bird was absent from the weathered granitic depression that served as a "crypt" and I, of course, was greatly disappointed. This rarely noted phenomenon of bird hibernation, together with Jaeger's contagious sensitivity regarding "Nature," instilled in me a conscious awareness and a lifelong enjoyment of birds.

While pursuing birds in the Southwest for five decades, I also researched the historic aspect of this subject, and another fascinating dimension began to unfold. The result of this endeavor is a chronological narrative about an assemblage of early explorers, naturalists, and scientists, who, along with their supporting eastern and European colleagues, joined in identifying and cataloging new southwestern avian discoveries from 1528 to 1900.

One of the great pleasures in writing this book is to acknowledge those individuals who have aided and encouraged me in the field of natural history. The conceptual idea to undertake the book was suggested to me by Shannon Davies, Editor of *Texas Birds*. I am grateful for her interest, advice, and guidance, which have continued during the course of this project.

I have also been fortunate to have a personal acquaintance with Gale Monson, with whom I have enjoyed several field exploits and who has provided me with many suggestions after reading the entire first draft. I am also indebted to R. Roy Johnson, who offered me his notes on early naturalists and, after reviewing the complete first draft, gave many helpful insights and comments. Joseph G. Hall kindly provided me with historic tapes and the closing epilogue quote by Loye Miller.

Many people have shared with me a common interest in natural history, especially birds, and have in their own way contributed to the presentation of this narrative. They include Jacqueline Adinaro, Al Anderson, Sandy Anderson, Josiah and Valaire Austin, Conrad Bahre, Al Bammann, Harriette Barker, Patrick H. Boles, Rick Bowers, Bill Broyles, Barbara Cain, Eugene Cardiff, George and Virginia Cechmanek, Kevin Cobble, Troy Corman, Alan Cox, Bill Davis, Larry Dembowski, Carol de Waard, Nancy Drew, George Drew, Rod Drewien, Dave Ellis, Jim Fairchild, John Fairchild, Jefford Francisco, William Gillespie, Richard Glinski, John Goodman, Grace Gregg, Kathy Groschupf, Drummond Hadley, Homer Hansen, Bruce Harris, Gayle Hartmann, Lupe Hendrickson, Don Higgins, John Higgins, Steve Hoffman, Jeanne Hopkins, Jim Horton, Bill Hoy, Tom Huals, David Jasper, Margaret Jensen, Carl and Jane Klug, Roy and Jessie Kniffen, "Dixie" Larkin, Dave Larson, Bill Litzinger, Larry Ludwig, Mary Magoffin, Matt Magoffin, Steve Marlatt, Wes Martin, Grace McKernan, Scott Mills, Arnold Moorhouse, Sonia Najera, Steven M. Norris, Tony and Thelma Nosek, C. J. Ralph, Don Ricker, Bill Rodgers, Ruth Russell, Steve Russell, Walt Saenger, Robert Schumacher, Gertrude Schwab, Chris Scott, Ted and Holly Scott, Helen Snyder, Noel Snyder, Pat Sonneborn, Sally Spofford, Walter Spofford, Richard Taylor, Steve West, Wayne Whaley, Alan Whalon, Jack Whetstone, and Sheri Williamson.

The Bureau of Land Management, Fish and Wildlife Service, Forest Service, and National Park Service, especially the past and present staff of the Chiricahua National Monument, have been helpful in numerous ways. Several institutions and libraries, such as the American Museum of Natural History in New York, National Archives in College Park, Maryland, Arizona Historical Society in Tucson, and the Universities of Arizona, California, and Texas, have been my primary sources for this project. I thank John Murphy and Roger Meyers, librarians and archivists, at the Special Collections Library of the University of Arizona; David Burgevin, museum specialist, and Jim Stead and Bill Cox, associate archivists, at the Smithsonian Institution Archives, Washington, D.C.; Mary W. Elings, pictorial archivist at the Bancroft Library, University of California, Berkeley; Carol M. Spawn, archives librarian, Ewell Sale Stewart Library at the Academy of Natural Sciences, Philadelphia; James J. White, curator of art at the Hunt Institute for Botanical Documentation, Carnegie Mellon University, Pittsburgh; the staff of the Museum of Vertebrate Zoology, University of California at Berkeley; Michele Poisson, Ayer Company, Publishers, Inc., Manchester, N.H.; Luke Shelby, New Mexico Department of Game and Fish; and Elizabeth Brooks, daughter-in-law of Allen Brooks. I also appreciate the photo approval given by Jeffrey S. Marks, managing editor of the *Auk,* and Walter D. Koenig and David S. Dobkin, current and past editors of the *Condor.* The invaluable service of interlibrary loan has linked me with at least fifty additional institutions when certain materials were difficult to obtain. The expeditious response by Colleen Crowlie, Kim Holman, and Ted Weller of the Cochise County Library was instrumental in helping secure numerous publications in this regard. The Tucson Audubon Society library was also helpful.

I am especially grateful to Christine R. Szuter, director of the University of Arizona Press, and the entire staff for their professionalism and high standards attained to complete this project. My appreciation also goes to two anonymous referees. I would also like to extend my gratitude to Joe Wilder for including this book in the Southwest Center Series. I am indebted to Heather S. Hopkins for carefully editing the manuscript and for writing this to me: "Adding to the experience is the fact that my 16-month-old son has just discovered birds. Every chance he gets, he goes to a window, points and says 'Birr' or chirps. I see him as a future ornithologist!" I thank Susan Alta Martin for her able assistance in computer-formatting the four maps.

I would also like to thank my children, Lisa and Mark, for their interest and the pleasure of their company when my wife, Charla, and I explored the southwestern landscape. The enthusiasm of Melissa and Stephanie, our granddaughters, has added greatly to the enjoyment of some adventures. The time-consuming writing and research of this endeavor has required patience and understanding by my wife, and she has also provided considerable assistance with the text.

Finally, I gratefully acknowledge the difficult, often dangerous, labors of the early naturalists as they left a fascinating history of discoveries, which would be unknown to us except through their rich trove of ornithological literature and bird collections.

Introduction

Among those who came to the Southwest during the nineteenth century were U.S. Army officers who recorded in detail their views, duties, and diverse experiences in this arid expanse. Partially from their perceptions, the notion developed that the landscape was foremost an inhospitable tract of venomous creatures, jumping cactus, blowing sand, endless drought, and intense heat.

These impressions have been conveyed in the form of numerous tales similar to those told by Dr. Elliott Coues, an army surgeon and naturalist, of his visit to Fort Yuma in 1865. Referring to the "excessive heat," he wrote "of the hens that laid hard boiled eggs" and "of the dog that ran howling across the parade ground on three legs because it burnt his paws" (Brodhead 1973). In 1870, while posted at old Camp Grant on Aravaipa Creek, Lt. John G. Bourke ([1891] 1971) of the Third Cavalry told of another aspect of military life by noting, "the story of the soldier who came back to Fort Yuma after his blankets, finding the next world too cold to suit him."

The times were difficult at best, and the keen observations, vivid descriptions, and amusing yarns given by the two young officers are understandable and hard to dispute. In 1853, prior to the visits of Coues and Bourke, another army surgeon and naturalist, Lt. Col. Thomas C. Henry, wrote of his experiences in New Mexico Territory: "This is a curious and unique country . . . full of hostile Indians . . . lizards, tarantulas, and flies in profusion" (Hume 1978). Continuing, he stated in simple but eloquent terms another facet about one of the major components of this book: there "are to be found many curious birds, peculiar to the country."

Since the time of Henry, over fourteen decades later, our many feelings, thoughts, and ideas about, and even fascinations with, the Southwest still vary considerably among all who reside in, visit, or read about the region. To those interested in natural history, and birds in particular, this area is one of continual stimulation and appeal.

In 1846, a two-year war began between the United States and Mexico. Subsequent to this conflict, the Treaty of Guadalupe Hidalgo was signed in 1848. The boundary was revised by the political compromise of 1850 and finally settled through the Gadsden Treaty in 1854. Final lines of the agreement were described by latitude and longitude or by existing physical or topographic features. The monuments marking this common border were generally made of tapered iron or stone monuments and today appear as white columns when observed from a distance.

The region delineated in this book lies generally between the 35th parallel and the irregularly positioned Mexican boundary. Within this zone, roughly 24° span the earth's surface in an east/west direction along the southern latitudes of North America. Beginning at the confluence of the Rio Grande with the Gulf of Mexico, the coverage extends westward into the Big Bend country of Texas through southern New Mexico and Arizona to the Pacific slope and islands of southern California. The immediate northern Mexican frontier and the Baja peninsula with its nearby islets are included, along with Guadalupe 160 miles further west.

In some instances the subject material is related to localities quite removed geographically from this general region. Among these exceptions are isolated bird discoveries that were made in either the more northern or southern latitudes but involve birds that also occur or migrate across these borderlands. An enormous landscape, the region comprises a wonderful diversity of interacting plant and animal communities.

This book is not intended to be a complete source or exhaustive treatise on all the birds of the region. It is, however, an effort to acquaint the reader with a facet not often discussed—the history of Southwest ornithology and the associated human record along the U.S./Mexico border. Our Southwest history, including the Indian presence, followed by the arrival of explorers, missionaries, settlers, soldiers, early naturalists, and scientists, has yielded a substantial record of contributions to our knowledge of numerous bird species. Indeed, this book is written with the expectation that those who delve into this subject will achieve greater gratification and erudition in a broader context about this history.

The accounts documented within this narrative form a trilogy: the wonderful variety of native birds, the early naturalists who discovered them, and the developing human history that brought them together. Nearly one hundred southwestern naturalists appear in the text, with at least an equal number of additional naturalists who are recognized because of their affiliation with specimens they received, studied, or named. Unfortunately, there are also gaps in our knowledge about this period that will never be filled.

SOURCES AND NOMENCLATURE

THE BASIC DOCUMENTATION for this book is mainly from published sources that have endured the test of time and appear in the large reference list. This literature is extensive and overwhelming. The major source for classification, nomenclature, and historic bird species documentation is the American Ornithologists' Union (AOU) *Check-List of North American Birds,* seventh edition (1998).

All scientific and common names of the birds included in the text follow the AOU *Check-List.* Subspecies are generally omitted. The current common name always appears in the text. To emphasize a particular point or in the context of a naturalist's quote, the scientific name is added. Both common and scientific names appear in the appendix or index.

Because of ongoing studies in ornithology, many birds are under constant taxonomic scrutiny, which sometimes results in nomenclature revisions. This may affect either, or both, the common or scientific names. For example, Elliott Coues (1878) referred to a "beautiful jet-black creature," later naming it with cumbersome embellishments "Crested Shining-black White-winged Fly-snapper." Today its common name is Phainopepla, an unusual singular form taken from the genus *Phainopela.* Normally, common names are descriptive in the vernacular and are usually binomial, but this bird is an exception. The origin for its common name was a result of William Swainson's scientific description of *Ptilogonys nitens* in 1838 and a subsequent genus revision to *Phainopepla niten* by Spencer Baird in 1858. The scientific name for this bird was derived from "Gr. *phainos,* 'shinning'; Gr. *peplos,* 'robe,' for the glitter of the plumage; L. *nitens,* 'shinning,' emphasizing the bright reflecting plumage" (Choate 1985).

Other examples and their respective reasons for change are noted throughout the text.

This book is not a complete repository for sequential nomenclature revisions. Citing repeated name revisions of particular birds presents a tremendous challenge and is useless verbiage to most readers. Consequently, many intermediate bird names will be skipped. However, the appendix, in addition to tracing bird name revisions dealing with eponyms of naturalists, includes a brief historical background for many others, including those appearing in the text. Common name changes have occurred much more frequently than scientific names. Specific name changes have rarely taken place, while genus name revisions have been frequent and are indicated by following the "format" of the AOU *Check-List*. If the bird species was originally described in another genus, the describer's name is shown in parentheses. The bird illustration citations follow a similar form.

Many of the bird collections to which these naturalists have contributed are housed in museums such as the Academy of Natural Sciences in Philadelphia, National Museum of Natural History in Washington, D.C., Carnegie Museum in Pittsburgh, Museum of Comparative Zoology at Harvard College in Cambridge, San Diego Natural History Museum, California Academy of Sciences in San Francisco, San Bernardino County Museum in Redlands, and the Western Foundation of Vertebrate Zoology in Camarillo, California. The universities and several state colleges of Texas, New Mexico, Arizona, and California, as well as those in Mexico, have extensive and important collections taken from the region. In addition, there are over forty collections by other institutions and individuals.

Four maps of the region have been compiled from numerous sources in the references. Not all the routes of the early naturalists are included; however, their travels correspond to most indicated locations with respect to time periods.

THE CURIOUS EXPLORERS

WESTERN EXPANSION ON the North American continent was motivated by a multitude of emotions from greed to altruism. These forces were nurtured by many human incidents that culminated during the 1840s and were verbally expressed as manifest destiny. This banner ultimately inspired Americans to acquire lands by force, or purchase, to the Pacific coast.

Prior to this demarche, however, many European countries continued to explore limited areas of mainly Spanish possessions which, in 1821, transferred to Mexico. This was a dangerous frontier of untapped material wealth for many and, for a few, a bountiful natural landscape yet to be explored, mapped, and cataloged for the benefit of science.

Amid the beleaguered native peoples and the determined nations struggling for possession of this land came a small group of inquisitive individuals. Regardless of nationality, they strove to observe, record, and reveal the natural features of this pristine expanse. Curious of mind, with a fascination for nature, they were the early naturalists, who, for well over a century and a half explored the Southwest and contributed to the study of its natural sciences. Many were army officers, surgeons, or both, with additional military responsibilities. Others were professional naturalists, while a few accompanied field parties without pay, or traveled alone. Often concerned with other obligations, many were also immersed in several aspects of the natural sciences. Although their ideology and struggles will be briefly discussed, the main focus of this narrative is their work with specific birds.

A majority of the early naturalists were active in the Southwest, while others, termed "closet naturalists," retired to the laboratory to study, paint, name, or classify specimens. Several were able to enjoy both aspects. Mentors, often fellow scientists, played significant roles by encouraging, financially supporting, or recommending them to accompany specific expeditions. In return, they often received specimens to study. In the late eighteenth and early nineteenth centuries, some naturalists were supported by various governments. And a few, as we shall see, did not fit into any of these molds.

Many individuals have been honored by having birds named for them, while others, for various reasons, have lost this recognition through subsequent nomenclature revisions. Being remembered in a patronymic form for a particular species should not be the only criterion for consideration as a prominent naturalist. Many of the people discussed in this book, therefore, have been included because they were commemorated in the naming of birds, while others are mentioned for their energies and devotion. The list is incomplete, with some individuals remaining obscure, or yet to be distinguished, and relegated to other bibliographies.

Although few of us have enjoyed the acquaintance of those involved historically in the study of birds, their legacy of literature has added to our pleasure, appreciation, and knowledge about this fascinating subject. The foundations of

INTRODUCTION

southwestern ornithology can be attributed to their diligent fieldwork, publications, and in some instances, teaching and institutional instruction. Through the experiences of these people, a great wealth of information has been added to this science.

The significance of their individual discoveries is, in most instances, difficult to measure or comprehend. Collectively, the study of birds, or ornithology, like all scientific disciplines, is a cumulative work of many people contributing to the sum of our knowledge. Much of their work still stands on its own merit. There are many examples of their forthright, often delightful, verse and prose as they interpret their thoughts of the period. However, those writings with an anthropomorphic approach, although appealing and often amusing, should be recognized to some extent for their literary rather than their scientific contributions.

Advancement of the natural sciences, in this case ornithology, has been rapid and orderly since the time of Carolus Linnaeus and Charles Darwin. It has developed in the Southwest through the earliest visitations of naturalists like William Gambel, James Abert, William Hutton, and the field associates of Spencer Baird, who among them include famous soldiers such as Elliott Coues and Charles Bendire. This progression of outstanding individuals terminates in this narrative at the close of the nineteenth century with renowned naturalists such as Henry Henshaw, C. Hart Merriam, Edward Nelson, Edgar Mearns, and Joseph Grinnell. The foregoing men are but part of a larger group of individuals, who merit closer consideration and examination.

It is unfortunate that in the limited space of this book, only brief glimpses of these people are viewed when looking back into their interesting and often rewarding lives. They were the pioneers, who explored an untrammeled land during a period of extreme hardship, uncertainty, and often hostility. They cataloged birds new to science and introduced them to the country and to the world.

The naturalists, as they appear, will follow with few exceptions in chronological order with respect to regional and important historical events. They will be introduced on the southwestern scene in a way not ordinarily found in conventional histories. As they search for birds, nests, and eggs, many specimens will be discovered that were new to science, while others will be added to our fauna north of the Mexican border. The viewpoints of one naturalist may corroborate or challenge another regarding behavior or identification of a species. Some of their comments may only be about observations and records. Still others may reflect something of their daily routine during this period of history.

Although most of the southwestern naturalists were working alone, they were most definitely affected by peers affiliated with eastern United States museums and institutions and by many other associates on the world scene. This international bond with science cannot be ignored. European systematists, most notably up to the mid-nineteenth century, greatly influenced the early descriptive science of North American ornithology. Together with the text, the appendix provides an overview of bird discoveries, past and present eponyms, and current nomenclature of most southwestern species encountered during this period.

The end of the nineteenth century marked a noticeable attitude shift by some of these people as they refined and advanced their science. Their disposition revealed an inconsolable distress at the continued human disruption on the landscape. Some began to express concern about diminishing bird numbers and the loss of individual species in this land of supposed resilient and potential fecundity.

This book, therefore, may reinforce for some the conviction that a perpetuation of a land ethic within our world is of such importance that it cannot be overstated. As the Southwest is enveloped by humans, birds caught up in this upheaval may be losers in the short term, while we will suffer their loss in the long term. We and those who follow will be sorely deprived of their appealing uniqueness. When we realize the human capacity for causing massive environmental change, we may see a direct correlation between the increased population of our kind and the shrinking natural world and its subsequent loss of biodiversity.

EARLY SOUTHWEST ORNITHOLOGISTS

1 THE EXCITEMENT OF DISCOVERY

"Scenes like these have little attraction for ordinary life. But to the Naturalist it is far otherwise; privations to him are cheaply purchased if he may but roam over the wild domain of primeval nature and behold 'Another Flora, of bolder hues / And richer sweets, beyond our garden's pride.'"
— THOMAS NUTTALL (Alden and Ifft 1943)

WITH THE APPEARANCE of the Spanish conquistadores in the Southwest came a prelude to recorded North American history, which included some of the first bird notes by Europeans. Many of the human disruptions that followed came about primarily through the explorations of men in search of treasure. Their ensuing actions also brought dramatic consequences to the unsuspecting native peoples throughout the region. During the initial Spanish conquest, however, other than geographic knowledge gained through navigation and a few vague notes on indigenous and captive birds, little was gained regarding the natural history of the region. Other explorers representing competing countries soon followed, effectuating significant discoveries.

EARLIEST ACCOUNTS, 1528–1777

FROM THE MISFORTUNE of a shipwreck on the coast of Florida and his subsequent removal to Follet's Island, southwest of Galveston Island, in 1528, Álvar Núñez Cabeza de Vaca (1993) began his terrestrial odyssey through parts of Texas, New Mexico, possibly southeastern Arizona, and northern Mexico. His

Relación indicated that he observed many birds, including "quail," and that the Indians from the southern regions traded "plumes and parrot's feathers" with others to the north for emeralds.

In the quest for the "Seven Cities of Cibola," Pedro de Castañeda, the official chronicler for the *entrada* (expedition) of Francisco Vasquez de Coronado (1510–1554), provided more early records. Although his notes were meager, they were among the first written accounts of birds (including "quail") given to explorers by the Indians. He also wrote of "a very large number of cranes and wild geese" and of the "cocks with great hanging chins." While at the pueblos, "tame eagles" were seen, along with dresses and long robes that were made from "fowles" that the Indians kept "for the sake of procuring the feathers" (Winship, 1892–1893).

In 1583, Antonio de Espejo, a Spanish merchant in search of three missing Franciscan fathers, mentioned that he had "found a parrot in a cage" when visiting the Pueblo Indians (Bolton 1916). Diego Pérez de Luxán, a member of the party, also recorded the apparent natural occurrence of the birds, while traveling through what is interpreted to be either Sycamore or Oak Creek in the San Francisco Mountain region south of the Grand Canyon. "At this place the river is surrounded by an abundance of grape-vines, many walnut and other trees. It is a warm land in which there are parrots" (Wetmore 1931). This may support the premise that the Thick-billed Parrot may have naturally occurred in this region at that time (Colton 1930; Hargrave 1939). Comparative osteology studies from numerous southwestern archaeological sites north of the present international border have also established the tropical Scarlet Macaw as a trade item and major source for ceremonial dress (Hargrave 1970).

In 1602, the barefoot Friar Antonio de la Ascension, aboard the *Santo Tomas,* second of two Spanish ships under the command of Sebastián Vizcaíno exploring the North American west coast, recorded in his diary a flock of vultures feeding on a beached whale near Monterey Bay. His notes indicated they were "the shape of turkeys, the largest I saw on this voyage. From the point of one wing to that of the other it was found to measure seventeen spans" (Harris 1941). During this period, an accepted span was eight inches. Thus the wing spread of eleven feet four inches, although slightly exceeding most modern measurements, confirms Ascension as the first European to observe the great California Condor. Along the west coast of Baja California, two prominent physical features bear the name of Vizcaíno: an expansive bay near the 28th parallel and, further inland, a striking desert of plains, hills, and mesas. Inhabiting the desert is a mixed forest of the unusual Cirio, a tall, slender, upward-tapering tree, and the giant Cardón cactus with massive arms.

The bird trade was again noted by Padre Luis Velarde in 1716 when he reported that "at San Xavier del Bac [Tucson] and neighboring rancherias there are many macaws, which the Pimas raise because of the beautiful feathers" (Wyllys 1931).

Half a century later, German Father Ignaz Pfefferkorn (1989), after his expulsion from the Southwest along with other Jesuits, entered in his account of colonial Sonora descriptions and habits of several birds. Among them, he carefully noted the *sinsontle* or Northern Mockingbird which "enchants the ear" and "mimics very cleverly the voices of other birds."

A decade later in 1777, Theodore de Croix, the new commandant general of the Interior Provinces, set out to inspect the vast region placed under his jurisdiction in New Spain. At the request of Croix, Father Fray Juan Agustín Morfi (1935), "under oath of obedience," accompanied him on this eight-hundred-league journey north into the region of what is now much of south Texas. During the six-month tour, Morfi detailed some of his observations on the little-known fauna by noting seasonal numbers of "ducks, wild geese, and cranes, of all kinds. . . . Along the banks of the streams and the outskirts of the woods the droves of wild turkeys are so numerous that they disturb the traveler with their clucking. The number of magpies, quails of all kinds, and wild hens, is incalculable."

These explorers and missionaries were still unaware of science with regard to the fauna they encountered. However, three decades earlier in 1741, German naturalist Georg Wilhelm Steller (1709–1746) aboard the *St. Peter,* a Russian ship under Danish-born Captain Commander Vitus Bering, became the first systematic collector of birds on the Pacific side of the continent when he landed in southeastern Alaska. On the Atlantic side, three English artist-naturalists had already illustrated many eastern birds. The first was John White in 1588, followed by *A New Voyage to Carolina. . . .* (1709) by John Lawson, and *The Natural History of Carolina, Florida, and the Bahama Islands* (1731–1743) by Mark Catesby (1682–1749). After the death of Catesby, two more Englishmen, George Edwards (1693–1773) and Benjamin White, continued the *Natural History* with separate editions in 1754 and 1771.

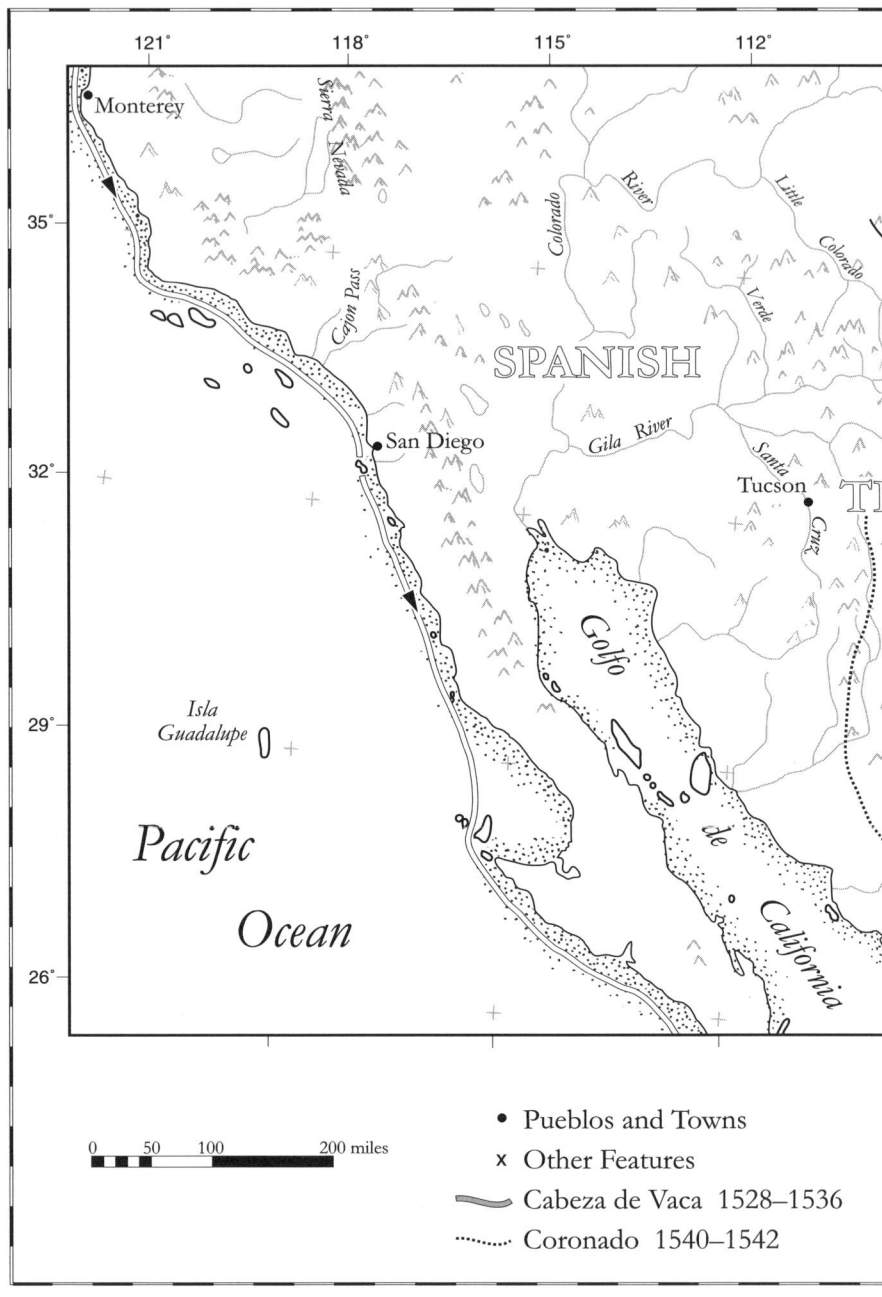

121° 118° 115° 112°

Monterey

Sierra Nevada

Colorado River

Little Colorado

35°

Cajon Pass

SPANISH

San Diego

Gila River

Santa Cruz

Tucson TE

32°

Golfo

29°

Isla Guadalupe

de

Pacific

California

Ocean

26°

0   50   100   200 miles

• Pueblos and Towns
x Other Features
⌇ Cabeza de Vaca 1528–1536
········ Coronado 1540–1542

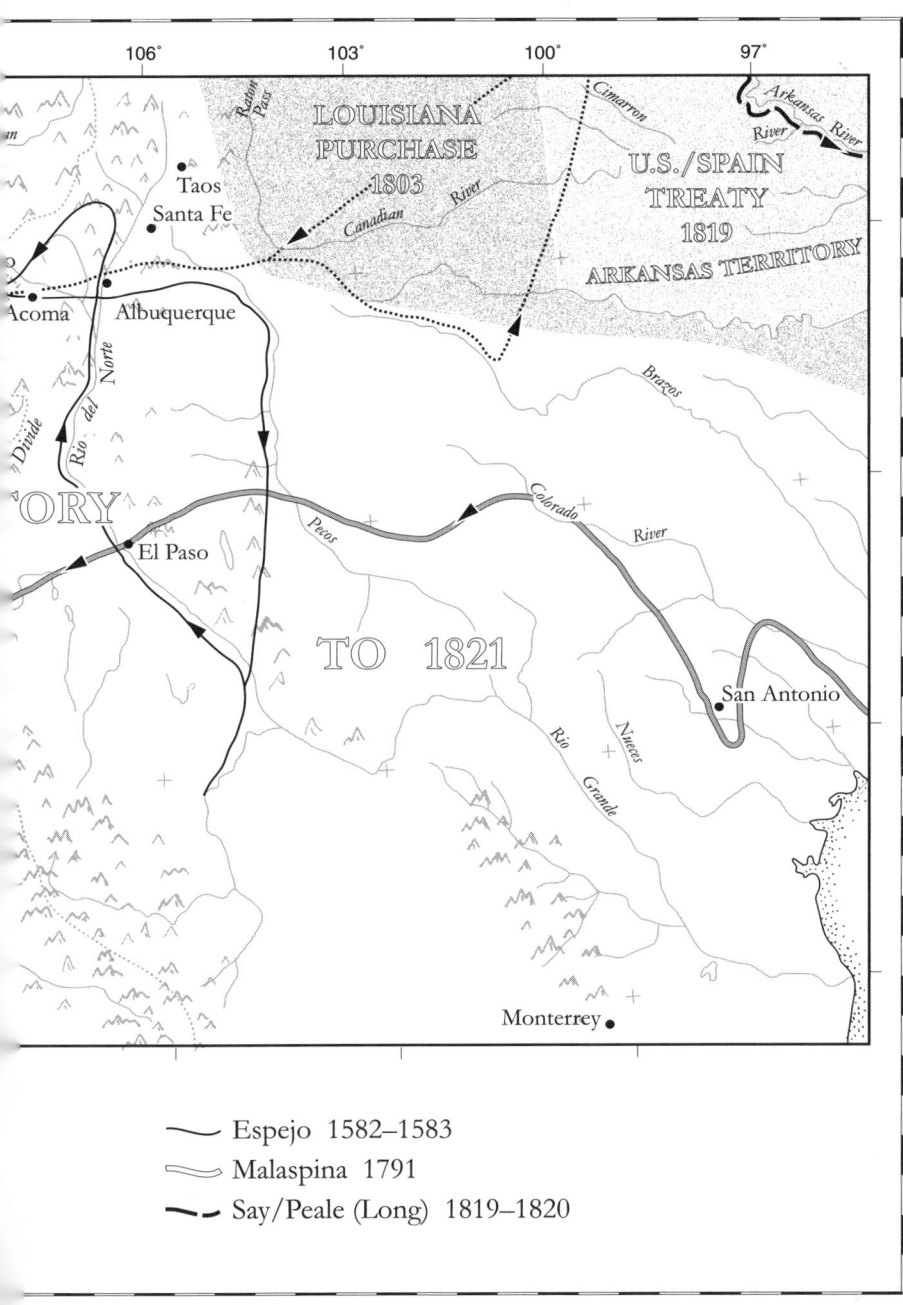

The Spanish Southwest, 1528–1821 (Map by Susan Alta Martin)

LA PÉROUSE'S EXPEDITION, 1786

ॐ⊚ SERIOUS INTEREST IN natural science began in the Spanish possessions in 1786 with the arrival of the ill-fated French brigs, the *Boussole* and the *Astrolabe,* to the Pacific coast under the command of Jean Francois Galaup de La Pérouse. Landing at Monterey, they collected two birds unknown to the world—"Perdrix de la Californie" and "Promerops de la Californie Septentrionale" (Alden and Ifft 1943). Fortunately, before their tragic disappearance, the scientists of this expedition dispatched home on another ship brief monographs, which were later included in an atlas featuring the two birds now known as California Quail and California Thrasher.

The quail was again collected and described, as we shall see, six years later. A half century passed before the thrasher was again found. It was described by William Gambel who named it *Toxostoma redivivum,* meaning "revived," because of its rediscovery after a long lapse of time.

LOST SPANISH TREASURES, 1787–1803

ॐ⊚ AT THE CLOSE of the eighteenth century, the Spanish conducted sixteen years of investigations into the natural sciences of New Spain. Shrouded in secrecy because of military concerns, their early contributions remained relatively unknown until well into the twentieth century. Referred to as the Royal Scientific (Botanical) Expedition to New Spain, the first group of explorers began their work in the vicinity of Mexico City in 1887. From this region they expanded their exploration to include the Spanish claims northward along the Pacific coast.

Several men are recognized as prominent naturalists in this venture. Among them were Spanish physician Martin de Sessé (1751–1808) and Mexican-born José Mariano Moziño (1757–1820), who extended his travels to the Pacific Northwest. The later phase of this enterprise was under the command of Peruvian-born Juan Francisco de la Bodega y Quadra and consisted of the vessels *Activa* and *Santa Gertrudis.*

The personnel of Spanish expeditions included scientific illustrators of great skill. Along with the work of Sessé and Moziño, their Mexican-born artists Vicente de la Cerda and Atanasio Echeverría (ca. 1772–?) deserve more than a casual glance. Several birds appear among the two thousand watercolors that were painted by the two artists (Engstrand 1981; Fleming 1983; Wilson 1970).

Their artwork was exceptional to the extent that Sessé wrote that Echeverría's painting of a butterfly was "so completely enchanting that it appeared to want to escape from the paper" (Engstrand 1981). Some question remains as to which artist painted particular birds; nevertheless, over two dozen had not been described up to this point, including California Quail, Greater Roadrunner, Blue-throated Hummingbird, and Yellow-headed Blackbird.

When Sessé and Moziño returned to Spain with the expedition documents, they found a climate of political turmoil in which their loyalties were questioned by the Spanish government. Much of their original work, including many of their paintings, remained unpublished at the time of their deaths. Except for some early copies, most of the original paintings disappeared until Jaime and Luís Torner of Barcelona realized the collection was part of their inheritance. These paintings, known as the Torner Collection, were then purchased in 1981 by the Hunt Institute for Botanical Documentation at Pittsburgh's Carnegie Mellon University (McVaugh 1982; Norris 1997).

During the intervening years, the four birds noted above were among others found almost immediately and described by other naturalists. The lack of prior published descriptions or available paintings of these birds left them open for succeeding naturalists to discover officially. Those naturalists will appear in future segments.

Another contingent of Spaniards, commanded by Italian-born Alejandro Malaspina (1754–?), led corvettes *Atrevida* and *Descubierta* down the Pacific coast, stopping at San Francisco and Monterey in 1791. While Malaspina listed many species of birds at Monterey, his talented artist, José Cardero (ca. 1768–?), illustrated several zoological plates, including a Northern Flicker. Like the work of other Spaniards of the period, most of these documents were not published, but remained in archival storage for over a century (Engstrand 1981).

ENGLISH DISCOVERIES, 1792–1831

FROM 1792 TO 1795, during the negotiations of the territorial dispute between England and Spain, Captain George Vancouver sailed the *Discovery* and *Chatham* along the California coast to explore and map the coastal waters above the 42nd parallel. Accompanying this voyage was Archibald Menzies (1754–1842), a Scottish naturalist who was on his second trip along western North America (Alden and Ifft 1943).

Greater Roadrunner,
Geococcyx californianus
(Lesson), 1829. The Spanish
Royal Scientific (Botanical)
Expedition of Sessé and
Moziño to New Spain,
1787–1803. (Courtesy of the
Hunt Institute for Botanical
Documentation, Carnegie
Mellon University)

Phasianus Cursor. Sp.N.
Hoitlallotl. Hern.ᶻ Mex. 25

Although Menzies collected two new birds, neither is named in his honor and he remains relatively obscure, probably because of his failure to describe either specimen. It wasn't until 1798, after his return to England, that Menzies' specimens were described by George Shaw. The first was the California Quail, previously noted, and the second was the California Condor.

From 1824 to 1827, and again from 1831 to 1834, another determined Scot, David Douglas (1798–1834), explored the west coast as far south as Monterey. Primarily a botanist, he is recognized by the common name of one of the most widespread conifers in the west, the Douglas-fir. Incidentally, this noble tree also bears the scientific name of *Pseudotsuga menziesii* in honor of its initial discoverer, Menzies. Like Menzies, Douglas (1829) also observed the great condor and, as a careful writer, he even noted in "Observations on the *Vultur Californianus* of Shaw" that their "quills are used by hunters as tubes for tobacco-pipes."

Tracing the travels of many of the early naturalists reveals bird observations that were noted in their journals, even though the birds were not collected to be described. An example is the Mountain Quail, which was collected and described by Douglas in 1826. The notes of Meriwether Lewis (1774–1809) and William Clark (1770–1838), however, disclose a description of this bird observed during their "Corps of Discovery" trek up the Missouri River to the Pacific Ocean and back from 1804 to 1806. Elliott Coues (1875), a great exponent favoring the priority of discoveries, cited them as the "real discoverers" and "original describers" by their "first acquaintance" with several animals since named by others. In noting their exploits, he stated they were not "trained naturalists, nor naturalists at all, excepting in so far as good observers in any new field, keenly alive to the requirements of the case, becoming naturalists as a matter of course."

Almost twenty years later, Coues (1893), even more eminently qualified, edited a *History of the Expedition under the Command of Lewis and Clark.* This edition revealed many insights regarding their abilities and collections. However, it was Paul Russell Cutright (1969) who, over seventy-five years later, portrayed them as "pioneering naturalists."

THE LONG EXPEDITION, 1819–1820

The first American military expedition west to include naturalists per se was to the southern Rocky Mountains in 1819 and 1820. It was under the command of Major Stephen H. Long (1784–1864), a topographic engineer, who included in his company Thomas Say (1787–1843). As one of the founding members of the Academy of Natural Sciences in Philadelphia, Say was known primarily as a systematic entomologist. His duties, however, were to "examine and describe any objects in Zoology" (Weiss and Ziegler 1931) along with his invaluable assistant, naturalist and artist Titian R. Peale (1799–1885).

Although it was nearly a disastrous venture, including several deaths and desertions, this was a rich expedition from the standpoint of avian discoveries (Thwaites 1905). It was a very trying trip, for Say endured many hardships. Not only did he suffer from ill health, but he was also robbed of his possessions and precious field note books by Pawnee and again by deserting soldiers.

In 1823, Say described the following nine new species that he and his party had discovered: Blue Grouse, Long-billed Dowitcher, Band-tailed Pigeon, Western Kingbird, Rock Wren, Orange-crowned Warbler, Lazuli Bunting, Lark

Sparrow, and Lesser Goldfinch. The specimens from the expedition were placed in Peale's Museum, later known as the Philadelphia Museum, which was originally opened in 1784 by Charles W. Peale, the father of Titian (Burns 1932). Titian's artwork, drawn while on the expedition, was also deposited at the museum.

It wasn't until after Say returned to Philadelphia and gained an acquaintance with Prince Charles Lucien Jules Laurent Bonaparte (1803–1857), the nephew of Napoleon, that two more new species in Say's collection were recognized. Bonaparte, considered by many to be the founder of American systematic ornithology, is honored by having the small, black-hooded Bonaparte's Gull named for him.

Bonaparte arrived from Europe in 1823, and while reviewing Say's collection, he discovered two undescribed birds. One was the Yellow-headed Blackbird. The second bird was a brownish flycatcher that Bonaparte named *Muscicapa saya* (Say's Phoebe). In doing so, Bonaparte (1825) wrote: "I dedicate it to my friend THOMAS SAY, a naturalist, of whom America may justly be proud, and whose talents and knowledge are only equalled by his modesty. The specimen now before us is a male, shot by Mr. T. Peale." A quarter of a century later he renamed Say's Phoebe *Sayornis saya,* which shares a generic name with the Black and Eastern Phoebes.

Of further interest regarding Bonaparte's immediate relatives is the generic name of *Zenaida* he applied to three members of the family Columbidae, two of which occur in the Southwest: the White-winged and Mourning Doves. They were given in honor of Princess Zenaide Charlotte Julie Bonaparte (1801–1854), niece of Napoleon, and cousin and wife of Charles.

The acceptance or disapproval of various taxonomists with regard to each other often surfaces in the literature of the period. Bonaparte, in his visit to Philadelphia, "set the whole Academy by the ears. He appeared to make warm friends and equally warm enemies" (Burns 1917). Elliott Coues (1879c), never one to be short in the praise or severe judgment of others, wrote another version of Bonaparte in very critical terms by noting his "schemes" of "coining . . . new names." In his censure, Coues felt the liberties and methods taken by the Prince in rearranging certain previously described species and generic designations were self-serving, "abominable," and a "scandal to science." Regardless, Bonaparte's contribution to American ornithology was significant, for he described a score of species, many of which occur in the Southwest.

Preceding Bonaparte was Scottish naturalist Alexander Wilson (1766–1813), who described at least twenty-five new birds and has eight North American birds named for him, more than any other naturalist. He wrote and illustrated *American Ornithology,* which was published in nine volumes from 1808 to 1814. Following Wilson's death, American-born George Ord (1781–1866) edited the eighth volume and then completed the ninth volume. In 1888, the Wilson Ornithological Society was named in his honor and its quarterly journal is the *Wilson Bulletin.* Bonaparte continued the work of Wilson by writing a four-volume work, *American Ornithology,* or *The Natural History of Birds Inhabiting the United States* (1825–1833).

ENGLISH MINING SPECULATORS, 1823–1824

✍ *I*N 1823, AFTER selling his private museum (London Museum), William Bullock (1775–1849) and his son, William, Jr., sailed for Mexico to invest in mining ventures. They collected many Mexican birds common to the southwestern border region. The same year, leaving his son behind to manage their mining affairs, Bullock returned to England with their specimens. English artist-naturalist William Swainson (1789–1855) described them. For his contributions to American ornithology, Swainson was later recognized by John James Audubon, Charles Bonaparte, and Thomas Nuttall, who each named a bird in his honor: Swainson's Warbler, Swainson's Hawk, and Swainson's Thrush. The first two birds also carry his name in their scientific names.

Although he never traveled to North America, Swainson was well known for his several publications and associations with American birds, describing no less than twenty species that extend their range across the northern Mexican border. He applied the geographic names of *mexicana* and *mexicanus* to several birds of that region. A Mexican discovery by Bullock that was described by Swainson illustrates one of several complicated nomenclature revisions which, in this case, occurred among three richly brilliant congeners over the past 150 years.

In 1973, Bullock's Oriole *(Icterus bullockii),* described by Swainson in 1827, was merged with the Baltimore Oriole *(I. galbula),* described by Carolus Linnaeus in 1758, under the name Northern Oriole *(I. galbula).* Then, in 1995, its original name was again applied. Some hybridization occurs between *bullockii* and *galbula,* but the gene flow is apparently restricted to a limited and stable zone in the Great Plains. Likewise, the Black-backed Oriole (*I. abeillei*), described by René Primevère Lesson in 1839 and named for the French collector M. Abeillé, although previously merged with *bullockii*, was again separated into a species. In understanding the justifications for the latest revisions, the AOU Committee on Classification reported in the 1995 supplement: "We know of no other case where there are so many discrete, abrupt, concordant differences between populations treated as one species."

A DISAPPOINTED NATURALIST, 1824–1838

IN 1824, THE aspiring young German naturalist Ferdinand Deppe (1794–1860) embarked on an expedition to central Mexico with his sponsor, a wealthy nobleman named Count von Sack, "Zweiter Ober-Jagermeister" and chamberlain to the King of Prussia. Shortly after arriving on the east coast, he separated from the Count and took to collecting with William Bullock, Jr., eventually meeting the senior Bullock. During a period of about two years, he assembled 958 bird skins representing about 315 species, and returned to Berlin (Stresemann 1954). Unable to gain an expected appointment to any of the scientific institutions in the city, he returned to Mexico with a botanist friend, Wilhelm Schiede, for more collecting in 1828. After this short-lived venture failed, he became associated with some merchants, which enabled him to travel and collect extensively along the Pacific coast to distant Monterey, California.

Most of Deppe's specimens were eventually delivered to Martin Hinrich Carl Lichtenstein (1780–1857), the director of the Zoological Museum of Berlin University. But Lichtenstein only gave names to the specimens, without giving descriptions. Among the several birds collected by Deppe and disposed of in this manner was the largest of the broad-winged raptors, the Ferruginous Hawk.

Lichtenstein provided limited notes of Deppe's rather dismal view of Baja California: "The peninsula is poor beyond description in living forms of organic life. Neither reptiles nor insects were seen anywhere" (Nelson 1966). Continuing, Lichtenstein did add that Deppe observed a "woodpecker" and that "there were only humming birds and a species of crow which indicated the presence of water in the vicinity." Edward W. Nelson (1966), a thorough naturalist who explored the region almost seventy years later in 1905 and 1906, took exception to Deppe's remarks: "The quotations . . . exemplify the kind of erroneous statements that gained currency during the early history of the penin-sula."

Many European scientists of the period who studied and bestowed names on Deppe's collection were Jean Cabanis (1816–1906), John Gould (1804–1881), Charles Bonaparte, Johann Wagler (1800–1832), and Philip L. Sclater (1829–1913). Although Deppe has remained relatively obscure, he, like the Bullocks, made significant discoveries of Mexican species, especially those that extend into the border region.

ENGLISH VOYAGERS, 1826–1827

ACCESS TO THE western slope of the continent by ship continued to be the main method of transportation for years to come. In 1826–1827, H.M.S. *Blossom,* sent by the Lords Commissioners of the Admiralty, visited western Mexico, Monterey, and San Francisco. Among the orders regarding the mission was the detailing of Captain Frederick William Beechey (1796–1856) solely "for the purpose and discovery of science." As Captain, he was to "afford the means of collecting rare and curious specimens" and at least two of each species were to be "reserved for public museums." The voyage was taken during a period of potential conflict with other foreign countries, although Spain had ceded all territories in the west to Mexico four years earlier. His instructions also stated, "In the event of England becoming involved in hostilities . . . you are not on any account to commit any hostile act" (Alden and Ifft 1943).

On board were three able men—surgeon Alexander Collie (?–1835), nat-uralist George Tradescant Lay, and Lieutenant Belcher—who, in addition to Beechey, collected specimens. Several birds were secured that were already known, but three new discoveries were made: Western Scrub-Jay, Pygmy Nut-hatch, and the California Towhee.

FRENCH TRADERS, 1827–1828

JUST MISSING THE *Blossom,* the French trading ship *Le Heros,* under command of Captain Auguste Bernard du Hautcilly, arrived on the west coast in 1827, where it visited many ports in California and Mexico (Palmer 1917a). On board was an Italian serving as ship's doctor, Paolo Emilio Botta (1802–1870). He collected several birds, and on his return home he gave two of them to Francois

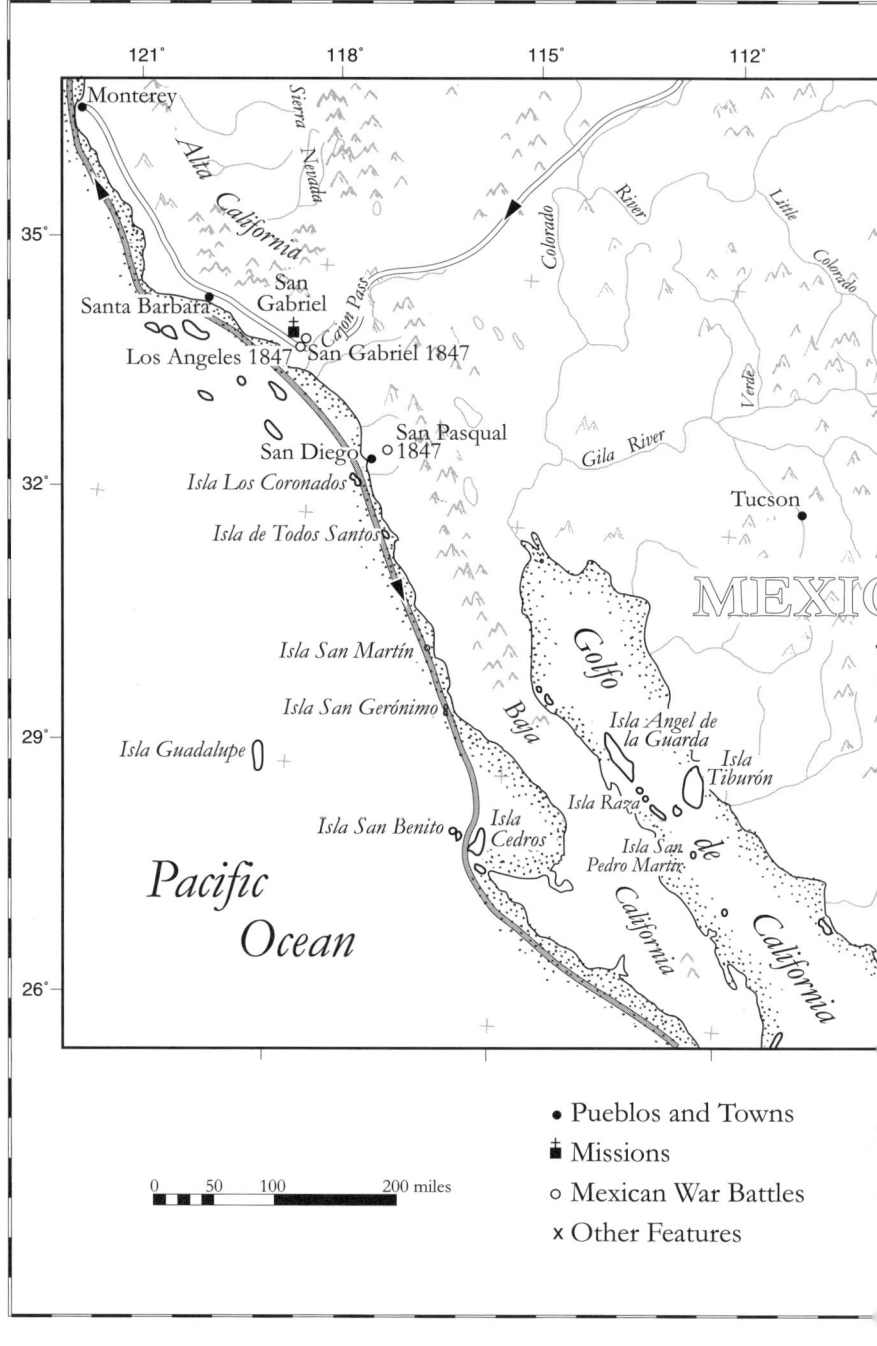

121° 118° 115° 112°

Monterey

Sierra Nevada

Alta California

35°

Santa Barbara

San Gabriel

Los Angeles 1847 San Gabriel 1847

Cajon Pass

Colorado

River

Little Colorado

32°

San Diego San Pasqual 1847

Isla Los Coronados

Isla de Todos Santos

Gila River

Verde

Tucson

MEXICO

Isla San Martín

Isla San Gerónimo

Golfo

Baja

Isla Guadalupe

Isla Angel de la Guarda

Isla Tiburón

Isla Raza

Isla San Benito Isla Cedros

Isla San Pedro Martír

29°

Pacific Ocean

California

de

California

26°

California

• Pueblos and Towns

✚ Missions

○ Mexican War Battles

✕ Other Features

0   50   100   200 miles

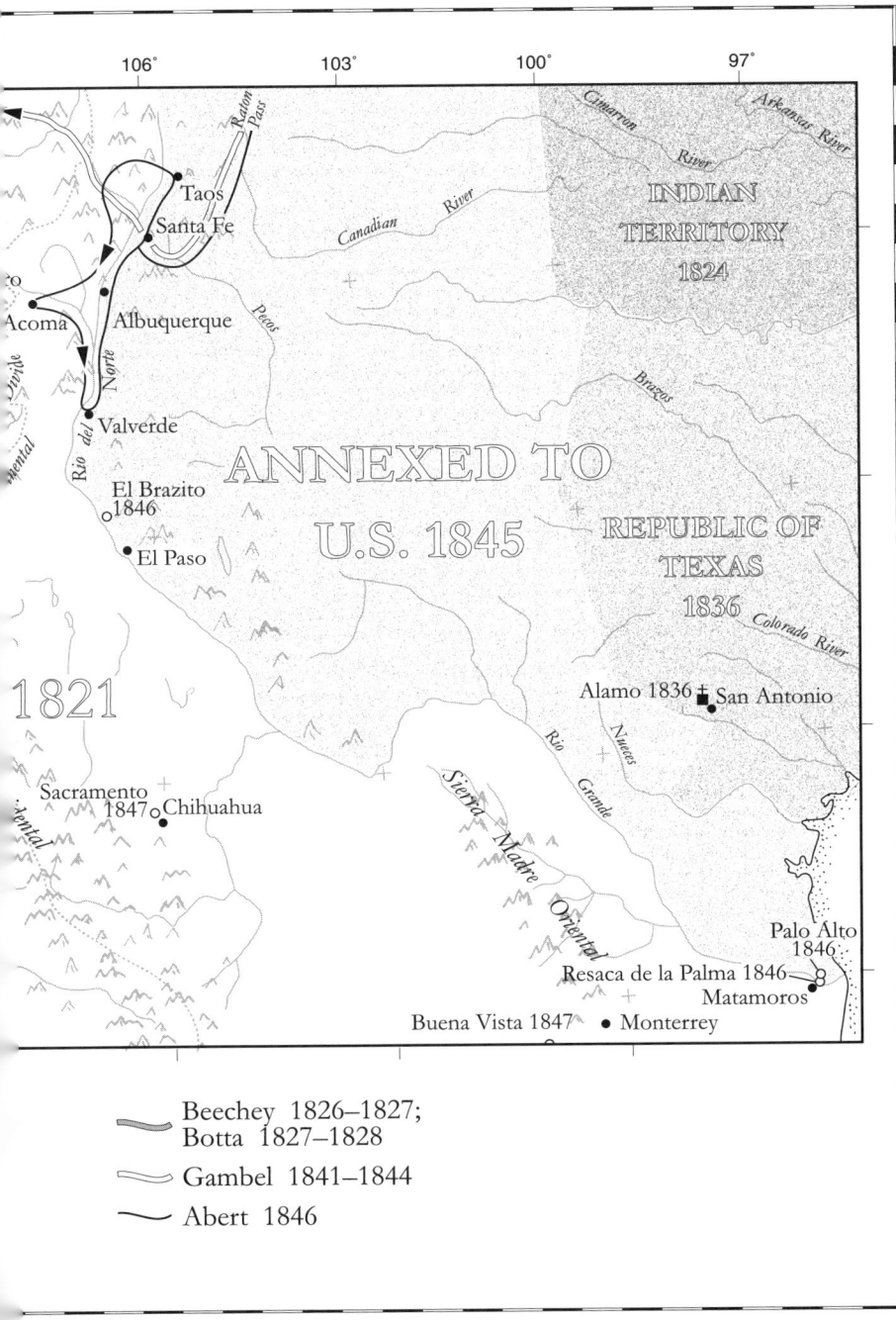

106° 103° 100° 97°

Raton Pass

Taos
Santa Fe

Cimarron River
Arkansas River

River

Canadian River

INDIAN
TERRITORY
1824

Acoma

Albuquerque

Pecos

Rio del Norte

Brazos

Valverde

ANNEXED TO
U.S. 1845

REPUBLIC OF
TEXAS
1836

El Brazito
1846

El Paso

Colorado River

1821

Alamo 1836 ☩ ■ ● San Antonio

Sacramento
1847 ○ Chihuahua

Sierra Madre Oriental

Rio Grande

Nueces

Palo Alto
1846

Resaca de la Palma 1846 ○
Matamoros

Buena Vista 1847 ● Monterrey

〜 Beechey 1826–1827;
    Botta 1827–1828
〜 Gambel 1841–1844
〜 Abert 1846

Mexico and Texas, 1822–1847 (Map by Susan Alta Martin)

Victor Masséna, Duke of Rivoli and Prince of Essling (1798–1863). In 1829, Masséna sent them to French naturalist René Primevère Lesson (1794–1849), who, in gratitude to Masséna, named one Anna's Hummingbird *(Calypte anna)* in honor of his wife, Anna Masséna, Duchess of Rivoli (1802–1887).

Botta wrote briefly about the habits of the second bird that Lesson described, which is sometimes affectionately called *el paisano* or "countryman" by the Mexicans or, north of the border, Greater Roadrunner. "The bird called charia runs very swiftly, jumping occasionally and beating its wings, which we might call flying. It is so poorly qualified to fly, however, that when it advances into the open it is possible, either on foot or on horseback, to catch it alive. It is known for destroying the rattlesnake and other reptiles." (Alden and Ifft 1943).

FRENCH WHALERS, 1837

🌀 IN LATE NOVEMBER of 1837, the French frigate *Venus,* on a three-year voyage around the world investigating whaling potential, anchored in Bahía Magdalena. This, incidentally, is the time when migrating Gray Whales return to this coastal region from their summer in the Arctic Ocean. The large lagoons and bays serve as protected nurseries and breeding areas for the great leviathans.

Other interests, however, caught the attention of the ship's doctor, Surgeon Major Adolphe Simon Néboux (1806–?). While ashore, he collected one of the tiniest members of the bird world, Costa's Hummingbird *(Calypte costae)* (Palmer 1918). Jules Bourcier, a French trochilid collector, described and named it in honor of a nobleman and serious collector, Louis Marie Pantaléon Costa, Marquis de Beau-Regard (1806–1864).

Although probably secured off the coast of Chile, another new bird found in the region was produced by the voyage: the Blue-footed Booby *(Sula nebouxii).* French ornithologist Alphonse Milne-Edwards named it for Néboux after he examined the collections from the expedition over forty years later in 1882.

AUDUBON'S CONTRIBUTION, 1827–1843

🌀 THE INTRODUCTION OF birds through beautiful art to the expanding population by John James Audubon (1785–1851) was indeed profound, and his discoveries and naming of new birds, although somewhat controversial, were of great significance. As a world famous artist and naturalist, his influences were not restricted to any particular territory or place on this continent. Although he never reached the arid Southwest, he painted and named many birds collected by others that occur in or pass through the region. Audubon applied names to ninety birds, although less than half were valid new forms. Of the remaining, most were previously named, several were not identifiable because of their descriptions, and a few proved to be hybrids. His friendships with people and the naming of birds for them even extended to several Europeans.

Audubon made two trips westward. The first was in 1837, when he traveled along the Gulf coast by ship to Galveston Bay and Houston. This expedition was taken during the time when his extraordinary epic, *Birds of America*, which was published in four volumes from 1827 to 1838, was being completed. The second adventure was in 1843 and lasted eight months. Starting in St. Louis, Audubon ascended the upper Missouri River to Fort Union, at the mouth of the Yellowstone River. Prior to these trips, Audubon realized that his opportunities to explore the West were rapidly diminishing. He anxiously awaited the results of a private expedition to the Northwest that left in 1834 under Captain Nathaniel J. Wyeth, of which two naturalists, John K. Townsend (1809–1851) and Thomas Nuttall (1786–1859), were members. After a successful trip, Townsend (1839b) decided to delay his return for another year, so he sent his collection home under the care of Nuttall, who arrived in Philadelphia in 1836.

Audubon, upon hearing of the arrival of several undescribed species, immediately sought their availability. He, after all, wished to include these in his last volume of *Birds of America*. His attempt was momentarily thwarted, however, by the "friends" of Townsend. "Having obtained access to the collection I turned over and over the new and rare species but he (Townsend) was absent . . . and loud murmurs were uttered by the *soidisant* friends of science, who objected to my seeing, much less portraying and describing, these valuable relics of birds, many of which had not yet been introduced into our fauna. The traveller's apetite is much increased by the knowledge of the distance which he has to tramp before he can obtain a meal; and with me the desire of obtaining the specimens in question increased in proportion to the difficulties that presented themselves" (Stone 1899a).

Audubon's zeal and determination halted only briefly, for he then enlisted the support of several others including Nuttall and Edward Harris (1799–1863), his close friend and benefactor. In an agreement with the Academy of Sciences in Philadelphia, with which Townsend was associated, Nuttall and Audubon published the new species under Townsend's name in the *Journal*. Audubon then purchased duplicate specimens through Harris for his subsequent use to describe and paint. Exacerbating the problem, Townsend (1837) fell into the awkward position of having Townsend's Warbler *(Dendroica townsendi)* named for him in the paper ostensibly authored by himself.

Unfortunately for Townsend, after a three-and-a-half-year absence and the success of his northwest trip disclosed, he was unable to publish his own work because of inadequate funds. He therefore sold his additional skins to Audubon and relinquished the recognition due him by contributing to the manuscript of Audubon's *Birds of America*. Townsend (1839a) did, however, compile another list of birds from the region.

Shortly after publishing the first volume of *Birds of America*, Audubon began work on his *Ornithological Biography* (1831–1839) with the help of the Scottish scientist, writer, and artist William MacGillivray (1770–1852). For Mac-Gillivray's efforts in editing and including technical data, Audubon named Mac-Gillivray's Warbler in his honor, but not without some annoyance to Townsend, its prior discoverer. Unknown to Audubon, Townsend (1839c), having gained employment with the Hudson's Bay Company, was also busy writing. He had described the same warbler two months earlier, honoring Dr. William Fraser Tolmie (1812–1886), his successor, after learning that Tolmie had collected the bird on Mt. Rainier. Audubon's application of MacGillivray for the scientific name was consequently voided for *Oporornis tolmie*. However, its common name is still retained.

In a science that continually undergoes revision, only four birds named in memory of Audubon remained so until the 1960s. Currently, the AOU lists just two: Audubon's Oriole and Audubon's Shearwater. Prior to 1979, the Yellow-throated Audubon's Warbler and the White-throated Yellow-rumped "Myrtle" Warbler were classified as two distinct species, until it was discovered that they sometimes hybridize freely in their overlapping ranges in the Northwest. Consequently the AOU lumped them as one species, the Yellow-rumped Warbler, and then recognized the western or "Audubon's" form as a subspecies, *Dendroica coronata auduboni*. The incipient distinctions of the two forms were obviously noted by the early naturalists, but they did not realize that the gradual divergences from isolation had not passed the point of completing separate speciation.

It is interesting to note that in the early application of eponyms, Swedish naturalist Carolus Linnaeus (1707–1778), the "father of biological taxonomy" who described over two hundred North American birds, has been completely overlooked in the naming process, for none commemorate him.

OLD CURIOUS, 1834–1836

🕭 BORN IN ENGLAND, Thomas Nuttall (1786–1859), one of the great field and laboratory naturalists, is worthy of further note for his great contribution to American botany and ornithology and for his brief visit to southern California.

Although Nuttall was considered frail by many, John Townsend (1839b), his field companion in the Northwest, obviously knew otherwise when he wrote of his enthusiasm and energy: "No difficulty, no danger, no fatigue has ever daunted him." Nuttall was so oblivious to surrounding dangers that he was sometimes prone to use his gun, muzzle first, for uprooting plants (Alden and Ifft 1943).

After taking two trips to the Sandwich (Hawaiian) Islands, Nuttall decided to leave for the east. In 1836, he traveled from Honolulu to Monterey and then down the Pacific coast to Santa Barbara, where he collected two birds restricted to the West. The following year, Audubon named the first of these birds the Tricolored Blackbird and the second, the Yellow-billed Magpie *(Pica nuttalli)*, for its collector.

In the beautiful, oak-covered foothills near Santa Barbara, Nuttall also observed the busy habits of Acorn Woodpeckers drilling countless holes in which they snugly placed acorns. Although this interesting woodpecker was discovered by Bullock in Mexico, Nuttall was first to note it north of Mexico.

After arriving in San Diego, Richard Dana, author of the famous sea narrative *Two Years Before the Mast,* noted his chance meeting of Nuttall strolling along the beach. They were about to share the *Alert* on their passage home. Both had been at Harvard: Dana as a student and Nuttall as a professor. Far from their previous common institutional setting, Dana (1964) described some of Nuttall's eccentric manners: "I was often amused to hear the sailors puzzled to know what to make of him, and to hear their conjectures about him and his business. . . . [they] called Mr. Nuttall 'Old Curious,' from his zeal for curiosities; and some of them said that he was crazy . . . to pick up shells and stones. . . . One of them . . . who had seen something more of the world ashore, set all to rights. . . . 'Oh, 'vast there! You don't know anything about them craft. I've seen them colleges and know the ropes. They keep all such things for cur'osities, and study 'em. . . . He'll carry all these things to the college. . . . Then, by and by, somebody else will go after more, and if they beat him he'll have to go again, or else give up his berth.'"

Two other western birds were named in Nuttall's honor: Nuttall's Woodpecker *(Picoides nuttallii)* described by Gambel in 1843, and Common Poorwill *(Phalaenoptilus nuttallii)* described by Audubon in 1844. Although Audubon is credited for discovering the goatsucker on the Missouri River, Captain Meriwether Lewis noted it less than one hundred miles upstream from the same location in 1804 (Cutright 1969).

Nuttall was not given to entering into competition or attaining popularity, but he had an insatiable thirst for knowledge of the natural sciences. He "was one of those rare beings so devoted to the cause of science that even the smallest personal comforts were discarded to further its purpose" (Alden and Ifft 1943).

Among the several books Nuttall authored was the two-volume *Manual of the Ornithology of the United States and Canada* (1832–1834) of which Elliott Coues (1878) remarked, "Nuttall like good wine does not deteriorate with age, and needs none of my brush." A revised edition in 1840 by Nuttall included many western species. He was a man highly respected in the natural sciences, with a considerable knowledge of ornithology, and the first American society devoted to birds was named the Nuttall Ornithological Club in 1873.

2 EXPANDING FRONTIERS

*"One object of our work is to present a general revision of the Ornithology
of the United States, endeavoring to bring our subject nearer to the true
state of the science than has been previously attempted in this country."*
—JOHN CASSIN (1856)

WITH THE TURBULENT year of 1836 began a period of bitter conflict
in key locations along the border region. The Mexican Army of General Santa
Anna stormed the old Spanish Mission of San Antonio de Valero. From this
bloody battle the rallying cry "Remember the Alamo" became symbolic to all
Texans. Later the same year, Sam Houston led an army that decisively defeated
Santa Anna at the battle of San Jacinto, thereby winning the independence of
Texas. The war between the United States and Mexico which followed enabled
growing numbers of settlers and miners to move westward into the recently
acquired territories.

ACADEMY OF NATURAL SCIENCES
OF PHILADELPHIA, 1812–

WITH THE THESIS of Linnaeus becoming the method of classification
and the accepted form of cataloging fauna, several established European museums, along with a few in the eastern states, began reviewing specimens. The
foundations for many scientific disciplines, including ornithology, were on sure
footing and in readiness for the exploration of the Southwest. During the first
half of the nineteenth century, most of the early ornithologists of this country

and those who were working on American birds abroad were either associated with, or communicated in some way with, the Academy of Natural Sciences.

Founded in 1812, the Museum rose rapidly to become a world-class institution. By 1825, it was, in many respects, surpassing Peale's Museum, which was eventually compelled to close in 1844. With its growing collection of specimens, the library became a repository of unmatched completeness, which enabled many men of science to study there.

Philip Lutley Sclater (1829–1913), secretary of the Zoological Society of London, a founder of the British Ornithologists' Union, and first editor of the *Ibis,* added his impressions about the library facilities after a visit to the Academy in 1856. "[T]he greatest liberality is shown to strangers who desire access," he wrote, and "there are few, if any, places in the globe where a student of Ornithology can pursue his researches with more convenience and profit" (Sclater 1857a).

With the expansion into the interior of the Southwest about to begin, the Academy was the only significant American museum to which naturalists could send specimens and by which they could have their works or findings published, in either the *Proceedings* or the *Journal.*

THE CASSINIAN PERIOD, 1842–1869

꘎ *H*AD IT NOT been for Dr. T. B. Wilson and the specimens he acquired for the Academy of Natural Sciences, John Cassin (1813–1869) would not have been afforded the opportunity to describe nearly two hundred birds from around the world—the most for any American (Stone 1899b). Although he acquired an interest in natural history at an early age, Cassin entered the import-export business. He did, however, make ornithology his avocation by becoming honorary curator of the Academy in 1842.

One of the most prominent figures of ornithology during his time, Cassin was considered by Audubon, his senior by almost three decades, to be a "closet," rather than a "field," naturalist. The two distinctions were apparent to all ornithologists of the period, and Cassin was decidedly attracted more to the gloomy laboratory and archives than to the adventures of the field. Cassin (1856) viewed the skill of bird collecting as the "ultimate refinement,—the *ne plus ultra* of all the sports of the field." But, he also concluded, the "very humble rank as a cultivator of Zoology" could only be achieved with "a combination of theoretical and practical acquirements . . . exactly in proportion to his scientific or systematic information," through "the museum and the library." It is through his laboratory work that Cassin is remembered as an influence on the ornithology of the Southwest.

In 1845, Cassin, in his first meeting with Audubon, considered Audubon "insufferable" after a disagreement over who was the first to discover the Harris's Hawk. Audubon named the raptor "*Falco harrisii*" for their common acquaintance Edward Harris. Unknown to the naturalist, the raptor had been named previously by Dutch ornithologist Coenraad J. Temminck (1778–1858) in 1824 (Morris 1902). Audubon's tendency to identify and prematurely name new species, like several other early naturalists, was deeply disturbing to Cassin.

Cassin became the first American taxonomist to study birds on a new level by using comparative research similar to what Europeans, like visiting systematist Charles Bonaparte, were using. This notable distinction was appraised by Elliott Coues when he wrote: "He [Cassin] was patient and laborious in the technic of his art, and full of book-learning in the history of his subject; with the result that the Cassinian period . . . is marked by its bookishness, by its breadth and scope in ornithology at large, and by the first decided change since Audubon in the aspect of the classification and nomenclature of the birds of our country" (Stone 1901).

As an author of numerous scientific papers and articles, Cassin took note of the westward expansion by publishing a book, which, in large measure, was on the Southwest, entitled *Illustrations of the Birds of California, Texas, Oregon, British and Russian America* (1856). Attempting to "supplement" the works of Audubon, Cassin also hired artists to paint the birds. Several southwestern naturalists supplied their field notes for use in the book. At least ten species of southwestern birds were described by Cassin as a result of his correspondence with the contributing field naturalists and Spencer Baird at the Smithsonian Institution.

The five birds named for Cassin, the most for an American-born ornithologist, were never seen by him in the field as they occur and breed in the West. They include Cassin's Auklet, Cassin's Kingbird, Cassin's Vireo (*Vireo cassinii*), Cassin's Sparrow (*Aimophila cassinii*), and Cassin's Finch (*Carpodacus cassinii*).

The practice of applying scientific names to birds has always been traditional, while instituting common English names came later. An example is when George Newbold Lawrence (1806–1895) introduced into usage Cassin's Kingbird a quarter of a century after its description by Swainson in 1826. Cassin

(1850) reciprocated by expressing in his description of Lawrence's Goldfinch (*Carduelis lawrencei*) that Lawrence was "a gentleman whose acquirements, especially in American Ornithology, entitle him to a high rank amongst naturalists, and for whom I have a particular respect, because, like myself, in the limited leisure allowed by the vexations and discouragements of commercial life, he is devoted to the more grateful pursuits of natural history."

Although interested in birds at an early age, Lawrence was thirty-five years old and working in the family New York mercantile business when he seriously became involved in ornithology through a meeting with John Graham Bell (1812–1899), a New York taxidermist, and Spencer Baird. Several southwestern birds were described by Lawrence (1851) in the *Annals of the Lyceum of Natural History of New York*. Cassin's respect for him was shared by Daniel Elliot, who wrote (1896): "With Lawrence ends an era of our science in the New World. . . . he belonged to the past, to the ranks of those who directed ornithological science into a new path."

Cassin's tireless affinity for laboratory work in classifying specimens and his constant exposure to arsenic undoubtedly contributed to his ill health and ultimately his death. Baird, who worked closely with Cassin for many years, exchanging specimens and thoughts, wrote they had "the warmest friendship I ever formed" (Dall 1915). It was twenty years after Cassin's death before another ornithologist, Witmer Stone (1866–1939), "a biographer of science and scientists," resumed work at the Academy. He became the major figure in organizing the Delaware Valley Ornithological Club which, in tribute to Cassin, designated their journal *Cassinia* in 1901 (Rehn 1941).

FORESTALLED BY A MALADY, 1841–1849

IN 1841, A young enthusiastic naturalist of nineteen from Philadelphia set out on an adventurous trip westward in search of new flora and fauna. He was William Gambel (1823–1849). Having been tutored under the famous Nuttall, Gambel felt amply qualified and excited about the long trip. From St. Louis, he left for Independence, where he proceeded with a wagon train in the company of traders and trappers to the Arkansas River, then through Raton Pass to Santa Fe, surviving two major skirmishes with Indians along the way.

Many new birds were discovered by Gambel during his short life. The first was a new chickadee that he found in the mountains west of Santa Fe. He later described and named it, unknowingly using a name previously given to another species: *montanus*. The duplication was corrected by Robert Ridgway, and the bird is now recognized as the Mountain Chickadee *(Poecile gambeli)*.

Continuing west Gambel crossed portions of two large deserts: the Great Basin, comprising intermountain plateaus, playas, and broad valleys, and the more southern Mojave Desert, named for a tribe of Indians living along the Colorado River. Gambel's discoveries were just beginning, for as he traversed this vast desert country dominated by small shrubs, he collected a new gallinaceous bird: "We met with small flocks of this handsome species . . . inhabiting the most barren brushy plains. . . . Here, where a person would suppose it to be impossible for any animal to subsist, they were seen running about in small flocks of five or six . . . uttering a low guttural call. . . . When flying they utter a loud sharp whistle, and conspicuously display the long crest" (1843). Gambel sent the notes to Nuttall, who not only published them in Gambel's name but also named the bird Gambel's Quail *(Callipepla gambelii)*.

On the coastal slopes of California, Gambel discovered three new bird species. In a willow thicket near Pueblo de Los Angeles, he collected and later named Nuttall's Woodpecker, honoring his close friend. Near Monterey, Gambel (1845) discovered a second bird, the gray-crested Oak Titmouse, "flitting actively about among the evergreen oaks of that vicinity in company with large flocks of several kindred species."

The elusive Wrentit, a unique North American bird representing a monotypic genus, was also discovered by Gambel (1845). Spencer Baird later noted it as a "very interesting species, which seems to combine within itself the principal characteristics of the Wren and Titmouse" (Baird, Brewer, and Ridgway 1874).

Gambel (1847) recorded his difficulties in finding this chaparral obligate. "For several months before discovering the bird, I chased among the fields of dead mustard stalks, the weedy margins of streams, low thickets and bushy places . . . [and heard] a continued loud, crepitant, grating scold, which . . . at last found it to proceed from this Wren-Tit, if it may be so called." Gambel continued his notes: "But if quietly watched, it may be seen, when searching for insects . . . jerking its long tail, and holding it erect like a Wren, which, with its short wings in such a position, it much resembles."

While in California, Gambel obtained a job as personal secretary to various succeeding U.S. Naval commanders. In that position, he had the opportunity to sail up and down the coast aboard naval ships. He also visited the same coastal mountain slopes where his famous teacher Nuttall had discovered the Yellow-billed Magpie in 1836. In his written observations, Gambel (1847) referred to the original name of "Nuttall's Magpie" given by Audubon. "I felt great pleasure on arriving at Santa Barbara . . . in seeing in its native haunts, the distinct and beautiful Magpie, discovered by my friend, the indefatigable naturalist and traveller after whom it is named; among others, a just tribute for the invaluable services he has rendered to natural science, during more than thirty years of his life."

Searching the old Franciscan missions built in the last half of the 1700s, he found the Barn Owl, the most widely distributed nocturnal bird in the world. Gambel's (1846) notes about this often misunderstood mouser follow: "This delicate feathered and familiar owl . . . I found very abundantly. . . . Its favorite resort is in the neighborhood of the towns and ruined Missions. . . . I have scarcely ever visited a Mission without disturbing some of these birds, which were roosting about the altar, chandelier, &c., of the chapel, and hearing the bendition [benediction] of the Padre for drinking all the oil out of the lamps. Every where in California, when speaking of it, we are sure to be told of its propensity for drinking the sacred oil; with what truth I cannot say."

While at the San Gabriel Mission, Gambel was the first southwestern naturalist to observe the Lewis's Woodpecker. Alexander Wilson first described it and gave it a preoccupied species name. This was corrected to *Melanerpes lewis* in honor of Captain Lewis by English naturalist George Robert Gray (1808–1872) thirty-eight years later. Gambel (1847) also observed the "peculiar" Clark's Nutcracker named for Captain Clark, the co-leader of the famed "Corps of Discovery."

In 1845, after Gambel returned to Philadelphia, Cassin excitedly wrote Baird: "Eureka! Gambel is here with his California birds and others—not very many, but some of the most magnificent specimens I ever saw" (Stone 1910). The young naturalist hurriedly went to work on his birds, even describing an auklet in honor of Cassin, not realizing it had already been collected by the Russians and described by German naturalist Peter Simon Pallas (1741–1811) in St. Petersburg in 1811.

This brought sharp criticism from Cassin about the methods and procedures of Gambel's laboratory work. "Gambel is exceedingly wild about describing, and it is already very difficult to get him to examine birds that he has concluded are new. . . . I mean in the woods of California without book—with scarcely knowing the names of late ornithologists. The birds that he has described are not examined . . . the most doubtful bird, too, probably at least, he has called after me" (Stone 1910). Gambel apparently had gained access to some of the offshore California islands to have obtained a specimen of this pelagic alcid.

In 1847, Gambel, failing to receive Cassin's support as curator at the Academy, set out again on another course in his life. The following year, he qualified to practice medicine, married, and decided to return ahead of his bride to California. It was then, after an extremely difficult December crossing of the snow-clad Sierra Nevada en route to San Francisco, that tragedy struck this twenty-six-year-old naturalist. Upon reaching the Rio de las Plumas (Feather River), he contracted typhoid fever while aiding sick miners. His death occurred shortly afterward and he was buried on Rose's Bar, a huge alluvial deposition along the river.

Gambel, like Nuttall, was greatly interested in botany and collected a wide variety of plants. Gambel Oak, a widespread deciduous tree of the southern Rocky Mountain region, was named in his honor. At the time of his death, his publications on California birds were the most complete for the west coastal region. What additional discoveries this young and energetic naturalist might have revealed to the field of ornithology will never be known.

ARMY OF THE WEST, 1846

NEVER IN THE annals of world history have the forces of the military been so involved in and contributed so much toward the natural sciences as in the American West and, indeed, the Southwest. Members of the U.S. Army in reconnaissance, conquest, and occupation were the leading edge in providing protection and preliminary natural history research in the region. Their participation continued from the first American Army party exploring west in 1819 until the end of the Apache Wars in 1886.

In the American way of developing ideas, there was a free interchange between the military and the various collecting institutions. Most of the officers were among the best-educated men of the period, and it was natural that their inquisitive desire to learn more about their surroundings often proved a benefit to science.

As the column moved westward, Lt. William H. Emory (1811–1887) wrote, "I looked in the direction of Bent's Fort, and saw a huge United States flag flowing to the breeze, and straining every fibre of an ash pole planted over the centre of a gate. The mystery was soon revealed by a column of dust to the east, advancing with about the velocity of a fast walking horse—it was 'the Army of the West.' I ordered my horses to be hitched up, and, as the column passed, took my place with the staff" (1848).

Lieutenant Emory was an officer in command of a small detachment of topographic engineers in the column commanded by Col. Stephen W. Kearny (1794–1848). The command was on a march from Fort Leavenworth, on the Missouri River, to "strike a blow at the northern provinces of Mexico which were New Mexico and California."

Assisting Lieutenant Emory was Lt. James William Abert (1820–1897), a graduate of the U.S. Military Academy at West Point and a topographical engineer. He had traveled as far as Bent's Fort the previous year under the command of Lieutenant John C. Fremont (1813–1890), "Pathmarker of the West." As Fremont continued west to California, he ordered Abert to remain and map the Canadian River from the Rocky Mountains downstream into eastern Oklahoma (Abert 1966).

With the opportunity to travel west again on a second trip, Abert (1848) was still not destined to reach California. His disappointment weighed heavily on his mind when he wrote on July 22, 1846: "I was taken ill, to such a degree that it was necessary to carry me in a wagon from that time until the 30th of July . . . my disease had obtained such an influence over my senses, that days and nights were passed in delirium, and a mental struggle to ascertain whether the impressions my mind received were true or false. Even my sight was affected, and when I gazed on Bent's fort, the buildings seemed completely metamorphosed. . . . The army under General Kearny marched on to Santa Fe, while I was left, harrassed with the thoughts of having come thus far, and having been stopped just as I was entering upon a field full of interest to the soldier, the archeologist, the historian, and the naturalist."

After recovering sufficiently, but missing the main command in Santa Fe, Abert was ordered by Lieutenant Emory, along with Lieutenant W. G. Peck, who was recovering from a similar sickness, "to remain for the present in the territory of New Mexico . . . [to] continue the survey of this territory." His travels extended across the "Rio del Norte" or Rio Grande to Laguna and the "Pueblos" where Coronado had searched for the fabled cities of "Cibolo." Here, he found the Indians with "some tame macaws, that must have come far from the south."

While the battles were being fought with the Mexicans in Chihuahua, it was still unsafe to venture far from the main river valley because of warring "Navajoes." Abert was continuing south slowly along the Rio del Norte beyond

Socorro, when he wrote on November 12: "On the way we saw several flocks of crested quails; that were running along with great rapidity . . . they seemed to depend more on their fleetness of foot than swiftness of wing. . . . The plumage is of a soft silvery grey, the iris hazel, and crest fringed with white." Abert was uncertain as to the quail he had collected, and its species determination was not fully understood until his return home.

On November 16, he observed, "we saw a fine bald-headed [Bald] eagle that was sitting on a bar in the middle of the Rio del Norte. We hailed it as an emblem of our victorious banner, which bears this bold bird on its folds." By November 21, the weather was becoming increasingly colder, causing him to remark the river ice "was sufficiently strong to bear the weight of a man."

On December 12, while camped at Valverde, Abert received orders "directing me to repair immediately to Washington city." Before leaving, he obtained five specimens of the quail, whose skins he "was obliged to fill . . . with corn meal" instead of the usual arsenic. His supply of arsenic, used as a preservative or insect repellent, had inadvertently gone to California with the troops and this had consequently restricted his collecting.

Returning up the river he "saw many cranes [Sandhill], the 'grulla' of the Mexicans, but they were not so numerous now as they were when we came down the river." Abert, endowed with many talents, displayed in his *Report* artistic impressions of them feeding along the "Rio de Norte."

Shortly after Abert's arrival home, his father, Col. John J. Abert (1788–1863), chief of the Topographical Corps, inspected the collection of birds taken on the trip. Ten years earlier, the colonel had assisted Audubon by procuring the cutter *Campell* for exploration in the Gulf of Mexico (Audubon 1960). Familiar with Audubon's work, but unable to find the previously mentioned quail in the painter's folio, Abert wrote at least four letters, including one on April 12, 1847, in which he wrote: "I considered the Quail a new one . . . unless it be out of place in the book and in that way has escaped my examination. A person of some knowledge in these matters, who has seen the skins, calls it a new bird, but there is no one of sufficient authority to depend upon" (Deane 1905).

At least one of Abert's quail specimens was received at the Academy of Natural Sciences along with a paper describing the "supposed new species." The review committee, however, recognized it as the bluish-gray Scaled Quail described by Irish naturalist Nicholas A. Vigors (1785–1840) in 1830. Although the quail was rejected as a new species, a portion of Lieutenant Abert's (1847) communication was accepted for publication. "These birds congregate in flocks or coveys of from 20 to 30 . . . and run with such rapidity, that it is difficult to get a second sight of a covey. . . . When forced to take wing, they make a whizzing sound . . . scattering as they fly, so that when they alight, they are at considerable intervals from each other, but they soon call together." John Cassin (1856) noted that Abert's publication was "the earliest record of this bird having been observed within the limits of the United States."

Like his engineer and naturalist father, who was recognized by the naming of Abert's Squirrel, James had Abert's Towhee *(Pipilo aberti)* named for him. Four years after Abert's report was published, Spencer Baird (1852), at the Smithsonian Institution, inserted in Captain Howard Stansbury's *Exploration and Survey of the Valley of the Great Salt Lake of Utah* the description of this new bird: "We have dedicated this species to its accomplished discoverer, Lieutenant Jas. W. Abert." Still, the circumstances surrounding the origin of the type specimen for this black-faced towhee remain doubtful.

John Davis (1951) gives one plausible explanation. Apparently, when reviewing Abert's private collection, Baird noticed this specimen was not among those noted in his trip west. As this species is absent from the Rio Grande drainage where Abert explored, but does occur about one hundred miles west in the upper Gila River valley, it seems uncertain that he collected the specimen at all. The complete story will never be known. However, it is assumed that Emory, or someone in the army, returning along the Gila River or the lower eastern desert riparian valleys of southern California, provided Abert with the specimen that ultimately reached Baird.

A TRANSPORTATION OFFICER, 1851

As a graduate of West Point, Captain Samuel Gibbs French served in southern New Mexico and west Texas, where he took an active part in the war with Mexico in 1846–1847. He was twice brevetted for gallant and meritorious conduct while serving in several conflicts, including those at Monterrey and Buena Vista, Mexico. His correspondence, and the specimens he sent to John Cassin, indicate that his interest in birds pertained primarily to the quails of the region.

French remained in the area at least until 1851, when he wrote of his experiences with the cryptic "Massena Partridge" between San Antonio and the Pecos River. "The wild, rocky hill-sides in the lone wilderness, seem to be their favorite resort; and there, where trees are almost unknown, and vegetation is scant, and where hardly a living thing is seen, are these fine birds found in all their beauty and gentleness . . . they would let a person approach within a few feet . . . and exhibit little of that wildness peculiar to all the other species of partridges with which I am acquainted" (Cassin 1856).

Since 1830, this Mexican quail, undeniably one of the most difficult birds to observe in the Southwest, has had its common name undergo a series of amendments. Its name has commemorated the French collector Victor Masséna (Massena Partridge) and army surgeon Edgar A. Mearns (Mearns' Quail). It was also called the Harlequin Quail, alluding to its unusual variegated plumage. This last comic epithet was then superseded by Montezuma Quail (*Cyrtonyx montezumae)*, which honored the Aztec leader Montezuma II (ca. 1480–1520) in the original scientific eponym.

In 1856, Captain French, like several officers of the period, resigned his commission from the United States Army and joined the Confederate States Army, later becoming a major general.

THE INSPECTOR-GENERAL, 1850–1852

꩜ A GRADUATE OF the U.S. Military Academy, Colonel George Archibald McCall served as an infantry officer in the war with Mexico, where he was twice brevetted for gallant and meritorious service in the battles of Palo Alto and Resaca in 1846. In 1850, after Mexico ceded much of its northern lands to the United States, New Mexico became a territory and California a state. During the following two years, McCall served as inspector-general of the army, traveling to the scattered military posts of the region.

McCall's travels in the region enabled him, as a naturalist, to view the bird life at hand. John Cassin (1856), with whom he frequently corresponded, remarked that McCall's "knowledge of the birds of Western America is unrivalled." Among the many birds he observed, two were added by him to our fauna in the lower Rio Grande region.

Most of the lower Rio Grande Valley of southern Texas is a delta or plain, which slopes gently southward into the Gulf of Mexico. The climate is semi-arid and subtropical and the plant distribution is often influenced by edaphic factors. While experiencing the characteristically dense, thorny Tamaulipan brushland, McCall wrote Cassin about the brilliantly colored Green Jay. "The first specimens of this Jay that I saw within the territory of the United States, were in the forest that border the Rio Grande, on the south-western frontier of Texas. There they were mated in the month of May, and no doubt had their nests in the extensive and almost impenetrable thickets . . . from the jealousy and pugnacity which they manifested on the approach or appearance of the large boat-tailed blackbird [Great-tailed Grackle] of that country . . . which was nesting in great numbers in the vicinity, I felt satisfied that the Jays were . . . also engaged in the duties of incubation and rearing their young" (Cassin 1856).

McCall also observed the only member of the tropical forest birds in the family Cracidae that extend northward to limited areas in the lower Rio Grande Valley. On hearing loud unfamiliar calls he wrote, "This very gallant-looking and spirited bird I saw, for the first time within our territory, in the extensive forest of *chaparral* which envelopes the *Resaca de la Palma,* a stream rendered famous in the history of our country by the victory achieved by the American forces under Gen. [Zachary] Taylor." McCall continued, "By the Mexicans it is called *Chiac-chia-lacca* [Plain Chachalaca], an Indian name, and doubtlessly derived from the peculiar cry of the bird. . . . I have observed a proud and stately fellow . . . mounting upon an old log or stump, commence his clear shrill cry" (Cassin 1856).

McCall (1851) traveled north through the open treeless upper Rio Grande and Las Vegas Valleys, where he occasionally observed diurnal Burrowing Owls standing on mounds in front of their burrows. "It was abroad at all hours of the day, and often amused me with its odd manners; taking wing at the near approach of my horse, and again alighting at the distance of a few yards, when it would face toward me and make, almost uniformly, three distinct and formal bows, with a mock solemnity that was irresistibly ludicrous."

Traveling west to the Colorado River, McCall inspected the isolated garrison at Fort Yuma on June 14, 1852. Continuing west he entered an extremely arid desert, called La Palma de la Mano de Dios, or "the hollow of God's hand," by the Mexicans. Because of its proximity to the Colorado River, it was named the Colorado Desert by geologist William Phipps Blake (1826–1910) the following year. McCall had entered into "a desert of sand of vast extent" when he arrived at "*Alamo mucho,*" a remnant cottonwood grove left from the previous meanderings of the Colorado River, but now "situated 44 miles" to the west and seventy miles east of "*Valle-cita* (little valley)."

As a man obviously concerned about great detail and gifted with keen observation skills, McCall described the "strikingly rich and full voice" of the Gambel's Quail and the seemingly inhospitable surroundings of the moment.

He was passing "the hours of noon stretched upon the sand . . . in the best shade obtained," enduring temperatures of "140° to 150° F." when he became aware of their presence. "A very good idea may be formed of his cry by slowly pronouncing, in a low tone, the syllables '*kaa-wale*,' '*kaa-wale*.' . . . There was to me something extremely plaintive in this simple lovesong, which I heard for the first time during a day of burning heat passed upon the desert. I had reached the well . . . before noon, and had halted to rest my jaded mules after their toilsome march. Here is, in truth, a desert!—figure to yourself, if you can, a portion of this fair earth, where, for some hundreds of miles, the whole crust seems to have been reduced to ashes by the action of internal fires" (Cassin 1856).

Continuing west into the low mountains near "*Valle-cita*," McCall approached a clump of trees, where he observed "a number of dark-colored birds [Phainopepla] . . . darting upwards from the topmost branches, and after diving and pitching about in the air for a moment, returning again to the dead branches with the lively port that proved them to be engaged in the agreeable pastime of taking their insect prey . . . they presented a pleasing sight, as three or four together were constantly either pitching upwards to a considerable height in the air, or gliding silently back to their perches. In these aerial evolutions, the bright spot on the wing . . . visible only when the wing is spread, gleamed conspicuously in the sunshine, and formed a fine contrast with the glossy black of the general plumage. I sat upon my horse, watching their movements for some time" (Cassin 1856).

Six years earlier, Lt. William H. Emory (1848), on his march along the Gila River to California with the "Army of the West," recorded his limited encounter with the same bird. "We secured to-day our long sought bird, the inhabitant of the mezquite, indigo blue plumage, with top knot and long tail. Its wings, when spread, showing a white ellipse."

TEXAS BORDER DETAIL, 1849–1853

AT THE CLOSE of the Mexican War, two army garrisons, established as frontier posts along the Rio Grande, were of major importance in the addition of several new species to the avifauna of the United States. Brownsville Barracks, situated on the Gulf of Mexico, was located eighteen miles upstream from the mouth of the Rio Grande amid an expanse of extensive prairies, sandy ridges, and shallow lagoons. Tracts of Mexican Palmettos prompted the early Spanish explorers to name this area of the Rio Grande, Río de las Palmas. Camp

Ringgold was approximately one hundred miles further up the Rio Grande, where several thornscrub species combine into dense thickets.

Among the first officers in the region to be recognized as a contributing ornithologist was Captain John Porter McCown (1815–1879) who, after graduating from the U.S. Military Academy, was brevetted for gallant and meritorious conduct at Cerro Gordo Pass, a major battle for Mexico City in 1847. As a field officer, McCown traveled through much of southern and western Texas. While posted at Camp Ringgold, he met John W. Audubon and witnessed the terrible tragedies of his California-bound party in 1849. The following year McCown was posted downstream at Brownsville Barracks.

Most of McCown's specimens were sent to George N. Lawrence (1851), who published his discoveries and observations in the *Annals of the Lyceum of Natural History of New York*. Among those were seven birds McCown added to our fauna north of the border: Black-bellied Whistling-Duck, Green Kingfisher, Vermilion Flycatcher, Verdin, Cactus Wren, Pyrrhuloxia, and Great-tailed Grackle. McCown also discovered three new species, which Lawrence described in the same paper. Two of them were the plain Olive Sparrow, which was collected in the scrubby thickets near Fort Brown, and the Ash-throated Flycatcher, which was secured between San Antonio and the Rio Grande.

The third discovery, McCown's Longspur *(Calcarius mccownii),* an inconspicuous flocking bird of open country that winters south into the central portions of the borderland region, is named in honor of this soldier-ornithologist. Lawrence, in writing its description, included a brief account of McCown's discovery. "Two specimens were obtained by Capt. McCown on the high prairies of Western Texas. . . . they were feeding in company with Shore [Horned] Larks. Although procured late in the spring, they still appear to be in their winter dress; in summer, I have no doubt they assume the gay and ornamented plumage of their congeners."

McCown's (1853) only paper written on Texas birds included habits of the Greater Roadrunner. "Often in my wanderings through the chaparrals on the Rio Grande, I observed piles of broken snail shells, and always near some hard substance, such as a bone or hard piece of wood. . . . I made many conjectures as to the probable animal. I never suspected a bird. . . . I heard at times . . . a sound very similar to that made by some woodpeckers by a rapid beating of their bill upon an old dry tree. This was also a mystery, as I could find no woodpeckers near the place the sound came from. Upon inquiry of a Mexican,

McCown's Longspur, Calcarius mccownii *(Lawrence), 1851.*
(Cassin 1856)

I was told that it was the Paisano breaking the snail shells to get at the snail . . . I was afterwards so fortunate as to see a bird so occupied."

The impending Civil War must have played heavily on the minds of the officers and men, as their attitudes, complicated by family and differing cultural backgrounds, aroused their individual emotions and loyalties. This was reflected, although in a very limited way, even among the early soldier-naturalists. Two prominent officers serving in the Southwest indicated their strong feelings. Major E. Coues (1878), an assertive surgeon naturalist, took exception to McCown's resignation and unchivalrous action of joining the Confederate States Army. Coues's subtle remarks were included with his comments about one of McCown's discoveries. "The Black-capped [tailed] Gnatcatcher . . . was discovered by Capt. J. P. McCown, then of the United States Army, who subsequently changed his allegiance to a temporary confederation which was declared in 1861."

Captain Charles E. Bendire, a greatly respected cavalry officer and a highly credible ornithologist, was usually quick to acknowledge most field ornithologists by noting their discoveries and general observations. With regard to McCown, however, Bendire only noted Lawrence's bird descriptions, and not the field participant, or collector. This is evident in his two-volume work *Life*

EXPANDING FRONTIERS

Histories of North American Birds (1892, 1895), where Bendire appears to have deliberately abstained from any inclusion of his one-time fellow officer's name, or personal achievements. McCown went on to serve as a major general in the Confederate States Army.

JOURNEY OF THE YOUNGER AUDUBON, 1849–1850

JOHN WOODHOUSE AUDUBON (1812–1862) began his early western trips with his celebrated father, John James Audubon, on an expedition to the Gulf of Mexico in 1837. The youngest son, John Woodhouse was, like his father, an artist who sought adventure. In 1845, with only one companion, he explored portions of the Texas wilderness. Then, on his final trip west, the younger Audubon undertook an almost disastrous journey from south Texas, through northern Mexico, southwestern Arizona, and on to California, writing of his experiences in *Audubon's Western Journal: 1849–1850* (1906). Much of the material regarding his experiences and long journey are taken from this source.

The decision of Audubon to undertake this western journey was prompted by the discovery of gold and "tales of speedily accumulated fortunes." Although he viewed the idea with "complete scepticism," he recalled the advice he had received from his "noble father": "Push on to California, you will find new animals and birds at every change in the formation of the country, and birds from Central America will delight you."

After joining a privately financed company in New York in February of 1849, Audubon traveled the interior to New Orleans, where he boarded the steamer *Globe* to Brazos, Texas. While Audubon and two companions went ashore and traveled by horseback, the main party proceeded aboard the U.S. government steamer *Mentoria* south along the coast and then up the Rio Grande to Brownsville. The two groups met at this point, where they boarded the *Corvette*, disembarking on the Mexican side just above Camp Ringgold, opposite Rio Grande City.

On the evening of March 12, everything seemed in order as the company grouped together on the banks of the river before continuing their overland journey into Mexico. Audubon wrote: "Gaily and cheerfully everything went on, under a clear sky like that of August at home, with all the soft, balmy, summer-like feeling. About me were the familiar notes of dozens of mockingbirds and thrushes. I opened out the nucleus of my collections, a little package of birdskins; a new thrush, a beautiful green jay, a new cardinal, were side by side

*John Woodhouse Audubon
(1812–1862). (Audubon
1906)*

with two new wood-peckers and a little dove, all new to our fauna, and I carefully spread them out to dry, and admired them."

Some of Audubon's specimens from the trip were sent to the Academy
of Sciences where Cassin (1856) credited him for collecting one of the "most
remarkable finches yet discovered in America." It was the handsome, striking
Black-throated Sparrow, which is mostly achromatic with colors ranging from
black, through shades of gray, to white.

Most of the higher mountain ranges of the Southwest enjoy more than
one species of charming birds represented by the genus *Baeolophus*, which
was designated by the German ornithologist Jean Cabanis (1816–1906) in 1850.
Audubon brought back a specimen of this genus, which was examined by
Cassin. It was similar to distinctly marked specimens sent by other collectors to

121° 118° 115° 112°

● Monterey

Sierra Nevada

GREAT BASIN DESERT

MOHAVE DESERT

NEW ME

Defia

CALIFORNIA 1850

35°

Colorado River

San Francisco ▲

Little Colorado

Tehachapi Pass ×
Tejon Pass ×

Santa Barbara ●

Los Angeles ●

× Cajon Pass

TERRITO

Santa Cruz Island

× San Gorgonio Pass

Santa Catalina Island

San Luis Rey ■

Warner's Pass
× Vallecito

Gila River

Pima Villages

San Pedro River

San Diego ■

Alamo Mucho Yuma

32° ┼

Isla Los Coronados

SONORAN DESERT

Tucson

San Xavier ■ Santa Rita

Isla de Todos Santos ○

Baboquivari ▲

Huachuca ▲

Nogales ▲

SONORA

Isla San Martín

TERRITORIO DE BAJA CALIFORNIA

Golfo

29°

Isla Guadalupe ○ ┼

Isla San Gerónimo

Isla Ángel de la Guarda

Isla Raza Isla Tiburón

Isla San Benito ○
Isla Cedros ○

de California

Pacific Ocean

Isla San Pedro Martir

26°

● Pueblos and Towns
✚ Missions
□ U.S. Army Garrisons
▲ Peaks or Mountains
× Other Features

0   50   100   200 miles

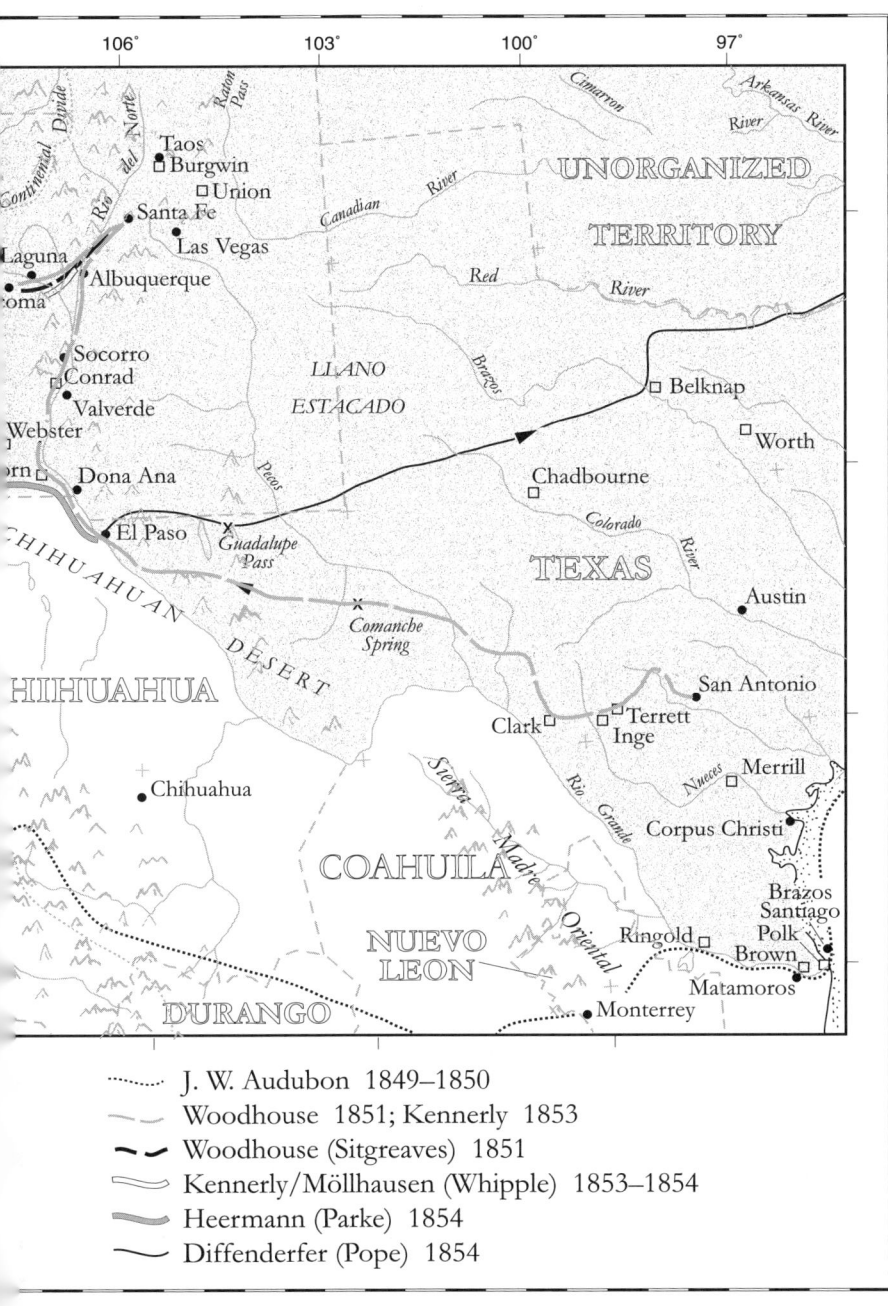

Expanding Frontiers, 1848–1854 (Map by Susan Alta Martin)

Map labels:

106° 103° 100° 97°

Cimarron
Arkansas River
River

Taos
Burgwin
Union
Santa Fe
Las Vegas
Albuquerque
Laguna
oma

UNORGANIZED TERRITORY

Canadian River
Red River

Socorro
Conrad
Valverde
Webster
Dona Ana
orn

LLANO ESTACADO

Pecos
Brazos

Belknap
Worth

Chadbourne

Colorado

El Paso
Guadalupe Pass

CHIHUAHUAN DESERT

Comanche Spring

TEXAS

Colorado River

Austin

CHIHUAHUA

Chihuahua

Sierra Madre Oriental

Clark
Terrett
Inge

San Antonio

Nueces
Merrill

Rio Grande

Corpus Christi

COAHUILA

NUEVO LEON

DURANGO

Ringold
Monterrey

Brazos Santiago
Polk
Brown
Matamoros

Legend:

.... J. W. Audubon 1849–1850
– – Woodhouse 1851; Kennerly 1853
⌐⌐ Woodhouse (Sitgreaves) 1851
≈ Kennerly/Möllhausen (Whipple) 1853–1854
～ Heermann (Parke) 1854
— Diffenderfer (Pope) 1854

Cabanis and Bonaparte. All three ornithologists described the bird in 1850, but it was Bonaparte's description of the Bridled Titmouse *(B. wollweberi)* that was retained because of a very close, but prior date. Little is known about Wollweber, a German who was credited for collecting it in Zacatecas, Mexico.

On March 13, the joyful bliss, high hopes, and expectations of a safe trip referred to on the previous evening were abruptly halted, for troubles developed in rapid succession. Almost immediately most of the company, including Audubon, became seriously ill with dreadful cholera. In short order, ten men died from the "horrors" of the intestinal disease. The group's troubles compounded when their money was stolen, their leader abandoned them, and twenty men decided to turn back.

While considering their plight, Audubon sought acquaintances among several "sympathetic" officers, including Captain John McCown, at nearby Camp Ringgold. McCown recommended sternly, "Go back, no one can do anything with volunteers, you have no power to compel obedience; now you go back honorably, and you don't know what you will have to endure on a march through Mexico" (Audubon (1906).

By April 28, the party had recovered from most of their difficulties and regained their spirits. Ignoring McCown's remarks, they set off for the interior of Mexico with Audubon elected as their leader. Their precarious position was stated by Audubon: "We are full two months behind our reckoning, and on a route of which I never approved, but which, when I took command, we were already compelled to pursue."

Once underway, they managed "twenty or twenty-five miles a day," depending on the availability of "water and forage for our horses," to Monterrey, crossing the Sierra Madre Oriental and the broad plains beyond. On August 22, after traversing the Sierra Madre Occidental, Audubon "left the coolness" of the mountains for the desert heat, crossing the Rio Yaqui into the great Sonoran Desert. Here, Audubon observed a familiar raptor. "[D]own we came to the table prairies and there were the Carra Carra Eagles in great numbers; sometimes we saw fifty in a day." The Crested Caracara, a large aberrant falcon, was painted by his father and formerly bore his name (Cassin 1865).

The younger Audubon continued north, crossing into the present state of Arizona along the west flank of a long ridge dominated by a central high granitic mass, Baboquivari Peak (7,730 feet). Passing through forests of Saguaro, Foothill Palo Verde, and Desert Ironwood, they arrived at the "Pimos" villages on the Gila River. Turning downstream, they endured sixteen days of extremely difficult travel, "and *such travel,* as please God, I trust we may none of us ever see again" (Audubon 1906). On October 14, after reaching Camp Calhoun (later named Camp Yuma), Audubon met with Lt. Amiel W. Whipple, who, as a U.S. Army topographical engineer, was establishing the new boundary between Mexico and California.

The struggling party then crossed "the much-heard-of and much-dreaded" Colorado Desert, arriving at Mission San Diego as "the last reflection of sunlight tipped the waves of the Pacific Ocean with gold" on November 3. On reaching the coast, Audubon renewed "old acquaintances" with seabirds such as the Brown Pelican, which he had observed in the Gulf of Mexico. The "wheels and plunges" for fish by this large remarkable bird added to his memories.

On November 7, he observed for the first time near Mission San Luis Rey a "California vulture [Condor], coming towards us until, at about fifty yards, having satisfied his curiosity, though not mine, he rose in majestic circles high above us, and with a sudden dash took a straight line, somewhat inclining downwards, towards the mountains across the valley and was lost to sight."

Although nearly one hundred men started the journey, only thirty-eight ultimately reached the gold fields of the Sierra Nevada. Audubon, after completing the arduous journey, returned east. Unfortunately, most of his valuable paintings and sketches left in San Francisco for later shipment went down with the steamer *Central America.*

EPONYM DIFFERENCES, 1847–1851

ॐ⊚ BESTOWING AN AVIAN eponym on a relatively unknown collector was difficult for a man like John Cassin, who was attempting to rectify past ornithological naming practices. The specimen for his consideration was collected by William Hutton while in Monterey on a visit to the Pacific coast during 1847 to 1851. Spencer Baird, however, from his office at the Smithsonian Institution, insisted in correspondence with Cassin that this collector be honored.

Cassin, in no uncertain terms, likewise expressed his views, "Calling that Vireo after your friend Hutton is one of the severest things. I don't want to do it—when he gets better known I will call something after him. This kind of thing is bad enough at the best, but to name a bird after a person utterly unknown is worse than that. I do not doubt his entire capability but I don't like to thrust honors upon him" (Stone 1916b). After a series of notes, Baird's position was finally upheld as Cassin later wrote: "Please give me the name in full of Mr. Hutton. . . . I must have it early as the paper goes to the printer." In his published description of Hutton's Vireo *(Vireo huttoni)*, Cassin (1851) acknowledged the collector as "Mr. Wm. Hutton, a zealous and talented young naturalist now resident at San Diego, in California, in honor of whom I have taken the liberty of naming it."

NEW YORK TAXIDERMIST, 1849–1850

THE SKILLS OF John Graham Bell (1812–1889) were well known by the leading ornithologists of the day, including John Cassin (1856), who recognized his abilities as "beyond comparison the most skilful preparer of birds and quadrupeds and general taxidermist in the United States." Working from his shop in New York City, Bell even instructed young Theodore Roosevelt (1858–1919) in the skills of taxidermy. Audubon selected Bell to serve as taxidermist on his memorable trip up the Missouri River and later named Bell's Vireo *(Vireo bellii)* in his honor.

Bell journeyed to California in 1849–1850 and collected several specimens. In Sonoma, Bell discovered the Sage Sparrow, which he also observed in San Diego. Cassin (1850) applied *Amphispiza belli* to it and wrote with the description that Bell was "a gentleman possessing a very extensive knowledge of natural history." Bell also returned east with two interesting new woodpeckers. The first, discovered near Sutter's Mill, was the White-headed Woodpecker, which ranges south to the mountains of the California and Mexican border.

The second was a sapsucker, the "Brown-headed Woodpecker *Melanerpes thyroideus*," which Cassin described in 1852. However, three years later, an avian serial about the identity of this bird began to unravel. John S. Newberry (1822–1892), a surgeon and naturalist serving on the Sacramento Valley to the Columbia River Railroad Survey, collected in southern Oregon a similar sapsucker but with an entirely different and more striking plumage. Newberry described this new discovery, naming it "*Picus williamsoni*" after Lt. Robert Williamson, the expedition leader.

The two sapsuckers remained distinct species for twenty-three years after Bell's initial discovery. Then, in 1873, Henry Henshaw (1875b; 1920a), collecting in the southern Rocky Mountains, discovered that sexual dimorphism is generally atypical among Picidae except for red markings on the head of the males.

Le Conte's Thrasher,
Toxostoma lecontei
Lawrence, 1851. Painted
by Spencer Baird.
(Baird 1859a)

By observing the two sapsuckers together in the field and then comparing specimens, he determined them to be dimorphic, in which Bell's skin was a female
and Newberry's a male. Appropriately, Williamson's Sapsucker *(Sphyrapicus
thyroideus)* honors the young officer, and the specific name is retained as Cassin
described it in 1852. Sexing specimens was obviously becoming an important
issue.

THE BEETLE COLLECTOR, 1850

In 1850, Dr. John Lawrence LeConte (1825–1883), a prominent entomologist and a specialist in Coleoptera, made a long trip in search of insects from
New York to San Francisco by way of Panama. He then continued down the

EXPANDING FRONTIERS

Pacific coast to San Diego, taking the usual road across the Colorado Desert to Camp Calhoun (Camp Yuma). While at the junction of the Colorado and Gila Rivers, he discovered a new, secretive dark-eyed thrasher. After securing a specimen, LeConte took the skin back to New York the following year. He gave it to George Lawrence, who named it Le Conte's Thrasher *(Toxostoma lecontei)*, commemorating its collector.

3 EXPLORATION OF THE CEDED MEXICAN TERRITORIES

"As the expedition will pass through the breeding-ground of many species of birds whose nidification [nesting habits] and eggs are not known, attention should be paid to securing abundant specimens of the nests and eggs. As far as possible the skin of the bird to which each set of eggs may belong should be secured, and have a mark attached common to it and the egg."
    —PROFESSOR S. F. BAIRD *(Stevens 1855)*

THE 1848 ACQUISITION of a vast tract of western lands by the United States from Mexico brought forth many problems concerning its retention, protection, and development. Two significant political developments quickly followed the ceding of these lands from Mexico and the Treaty of Guadalupe Hidalgo. In 1849, gold was discovered in the Sierra Nevada and, primarily because of rapid population growth, California was accepted into the Union as a state in 1850. Lagging behind in the statehood process, the remaining southern portion of lands covered by the Treaty, from Texas to the Colorado River, were recognized as the New Mexico Territory the same year.

THE TOPOGRAPHICAL ENGINEERS

THE EXPLORATION OF this huge landscape for military and civilian reasons required a massive undertaking. The awesome task naturally fell on the Corps of Topographical Engineers which, created in 1838, had men with considerable experience. Under the command of Col. John James Abert, its personnel consisted primarily of the previous topographical engineers, established during the War of 1812.

Although essentially a military unit, the officers within the Corps were an elite group comprising mostly graduates of the U.S. Military Academy at West Point, the premier school of engineering in the country. Beginning in 1802, the studies at this school afforded not only an education in engineering, but opportunities for acquaintances and contacts with the leading scientists of the day.

In exploring the West, the engineers were to map, locate boundaries, and determine the best and most economical direct routes to the Pacific coast. Each expedition required the principal men to be capable and proficient in the use of transits, theodolites, sextants, barometers, and chronometers. Horizontal and vertical positions, often geodetic, were determined on the ground by traversing or by triangulation techniques initiated from astronomical observations.

Several engineers closely associated with the early naturalists were engaged in many military clashes or tested on the field of battle. Lt. William H. Emory was beside Gen. Stephen W. Kearney, commander of the conquering "Army of the West," when he was severely wounded in the charge at San Pasqual near the approach to San Diego. Capt. William H. Warner, Lt. John W. Gunnison, and Lt. Amiel W. Whipple (1859, 1853, and 1863, respectively) were mortally wounded in separate engagements.

In 1849, Colonel Abert expressed concern that the "integrity of the Union" had to be maintained through direct communication and transportation routes extending to the West. "Unless some easy, cheap, and rapid means of communicating with these distant provinces be accomplished, there is danger, great danger, that they will not constitute parts of our Union" (Goetzmann 1991). His position was ultimately supported and funded by Congress for the formation of several surveys.

THE ASSISTANT SURGEONS

WITH THE CORPS of Topographical Engineers providing a framework of organization during much of the initial exploration, other elements of the army, namely the men of the Medical Corps, provided much of the fieldwork in investigating and collecting natural history specimens. Most of these men had a dual role of being surgeons as well as naturalists and carried out both successfully.

During this period a rank of first lieutenant or captain held the position of assistant surgeon, and surgeon was attained with the rank of major. Most of the doctors in the West did not achieve a higher rank than this while in the field, so the designations to surgeon general will not be discussed. A few doctors were also civilians, who, for one reason or other, could not, or did not, enter the Medical Corps, but were designated as acting assistant or contract surgeons. The need for surgeon-naturalists, regardless of status, to accompany the expeditions was recognized and demonstrated by the cooperation given throughout the War Department, from Secretaries Charles M. Conrad and Jefferson Davis, down through the commands.

ZUNI AND COLORADO RIVER EXPEDITION, 1851

АT THE CLOSE of the war, Colonel Abert submitted to the Academy of Natural Sciences of Philadelphia a request for a surgeon and naturalist to join a boundary survey in Indian Territory (Stone 1904). Dr. Samuel Washington Woodhouse (1821–1904) was recommended and subsequently appointed acting assistant surgeon to serve under the command of Lt. Lorenzo Sitgreaves (1810–1888) in 1849.

Sitgreaves was next directed to command on a reconnaissance that was to form in Santa Fe and explore possibilities for a wagon route to California. Serving again under Sitgreaves, Woodhouse (1853) included an account of their journey in a *Report of an Expedition down the Zuni and the Colorado Rivers in 1851*. Captain Sitgreaves is remembered for a National Forest and a mountain (9,388 feet) south of the Grand Canyon that bear his name.

The route that Woodhouse took to Santa Fe was not the most direct, for he went from San Antonio to El Paso and then up the Rio Grande. After reaching San Antonio, however, he "suffered much from intermittent fever" to the extent that he was limited "in the pursuit of [his] favorite studies" as a naturalist. In spite of this limitation, and even before the official part of the expedition began, this first leg of travel in southwestern Texas proved to be of special interest.

It was in April, "on the prairie near San Antonio," that Woodhouse collected a small bird that he at first mistook for a Savannah Sparrow, which "it much resembled in its habits, but upon examination it proved to be totally distinct." On his return to Philadelphia, Woodhouse (1852a) named the bird Cassin's Sparrow *(Aimophila cassinii)* "in honor of my friend Mr. John Cassin, the Corresponding Secretary of the Society, to whose indefatigable labor in the department of Ornithology we are so much indebted."

In May, Woodhouse, traveling "two hundred and eight miles west of San Antonio" to the Rio San Pedro, was "attracted" to a bird giving a "singular note" that was "feeding in the dense cedars." After some difficulty, he "procured two specimens, both of which, on dissection, proved to be males." His discovery was the Black-capped Vireo which, unknown to him at the time, proved to be dimorphic. Characteristic of this species, but unusual for this genus, the glossy black on the head of the male is replaced by slate gray in the female. In 1878, W. H. Werner discovered the first nest of this rare vireo along the Guadalupe River in Texas (Brewster 1879a), and painted a lovely watercolor depicting a pair of the "Greenlets with their nest" (Coues 1879a).

On August 15, 1851, with Lieutenant Sitgreaves in command and Lt. John G. Parke as assistant, both topographic engineers, the expedition began their march from Santa Fe west to Los Angeles. Crossing the "Rio de Norte" and moving up the Rio Puerco to Laguna, Woodhouse collected from among the pinyon-juniper woodlands a specimen that the men referred to as a "pinon bird." Although known earlier as "Woodhouse's Jay" in this region and "California Jay" on the Pacific slope, this corvid was renamed Western Scrub-Jay in 1995.

Continuing westerly, they came upon a huge rock face of soft sandstone with numerous inscriptions made by a succession of passing travelers. El Morro, standing as a noteworthy geologic feature, also serves as a historical marker. The first non-Indian to inscribe the rock face was a Spanish conquistador, Governor-General Don Juan de Onate, on April 16, 1605. Six years earlier, he had passed this place and drunk water at its base, naming it "Aqua de la Pena"— Water of the Rock. In 1849, Lt. James H. Simpson (?–1883), another topographic engineer, who was engaged in an action against the Navajo Indians, was the first American to examine it and report its existence.

From an ornithological standpoint, this picturesque location marks the discovery of a new, interesting, and somewhat controversial bird species. Woodhouse (1853), in his preface remarks of the expedition, wrote: "Encamped at the Inscription Rock, a singular sandstone mesa about two hundred and fifty feet high. Here I observed a new swift, of which, however, I was unable to secure a specimen, but I was close enough to become well acquainted with it; I propose for it the name of the Rock Swift (*Acanthylis saxatilis*)."

Woodhouse went on to describe his actions: "Being on the top of this high rock at the time without my gun, I was unable to procure specimens. . . . I descended immediately and procured my gun; but the birds by this time flew too high. . . . They were breeding in the crevices of the rock . . . it being the only place that I have observed them."

Woodhouse never received recognition for the visual discovery of the White-throated Swift from either John Cassin or Spencer Baird, the two leading taxonomists of the period. Both men were aware of the incident, but because Woodhouse lacked a specimen, the identification of the proposed new bird, even though published, came into question and was ignored. In 1854, this species was collected three hundred miles further west by Caleb Kennerly and Baldwin Möllhausen. Their specimen was described by Baird (1854), but it took seventy-seven long years before a new genus name included the specific name first applied by Woodhouse to read *Aeronautes saxatalis*. Several succeeding early naturalists, including Kennerly, John Xántus, Elliott Coues, and Robert Shufeldt, commented on Woodhouse's encounter with the swift. Their respective remarks will be noted as these men appear in the text.

On September 1, thirty miles west of El Morro, they arrived at Acoma, a Zuni pueblo that served as the actual commencement point for the expedition. Here, while on the Continental Divide, the party was detained for over three weeks as the army engaged in a punitive action against the Navajo Indians. The delay was costly for Woodhouse, for in an attempt to collect a rattlesnake, he was bitten on the left index finger. In addition to the pain, and loss of part of his finger and nail, it "was a sad accident for me," he recalled, "more particularly at this time, as we were just about commencing the most important and interesting part of the exploration. I did not recover the use of my left hand for months afterwards, and this accounts for the small collection of birds."

Resuming their march, the party passed "agatized" (petrified) trees and the Grand Falls on the Little Colorado River. They crossed broad plains with lava flows and cinder cones as they approached the conically shaped San Francisco Peaks (12,633 feet). During the second week of October, they entered the mountains, where Sitgreaves (1853) wrote of the "beauty in some of the glades and mountain glens" and of the "aspens of a brilliant yellow." In summer, darkened forests extend to the bare alpine summits. The scene transforms mid-mountain into a ringed brilliance of yellow aspen in fall, and then climaxes in a winter finale of mountains engulfed in glimmering snow. From these mountains, Woodhouse added the liver-colored Hepatic Tanager to our avifauna north of Mexico.

While in the San Francisco Peaks region, Woodhouse (1852b) was the first naturalist to encounter a "beautiful, large, and tufted-eared squirrel," which feeds primarily on the bark, buds, flowers, and seeds of Ponderosa Pines. After revising its specific name because it was already "applied," he renamed it in honor of his mentor. "I propose now to call it SCIURUS ABERTI, after Colonel J. J. Abert, chief of the corps of Topographical Engineers, U.S. Army, to whose exertions science is much indebted." The Colorado River divides the tassel-eared Abert's Squirrel into two forms; those on the north rim are very dark in color with a white bushy tail (Kaibab), while those on the south rim, the type Woodhouse collected, are more uniformly gray.

Their trip to the Colorado River and then south to Yuma was over extremely difficult terrain "much broken by precipitous ravines" (Sitgreaves 1853). The party suffered from lack of food and water and many pack animals gave out. The Indians they encountered, such as Yampai (Hualapai), Mohave, and Yuma, all attempted to impede their progress. Their guide, Antoine Leroux, was seriously wounded by arrows in an ambush, and a lagging soldier was killed in another attack. After reaching the Colorado River, Woodhouse was hit in the leg by an arrow while warming himself before a morning fire. On November 30, the exhausted party finally reached Camp Yuma at the mouth of the Gila River, where their provisions were replenished for their final push to California.

THE BAIRDIAN PERIOD, 1850–1887

THE MID-CENTURY PROBABLY marked the most important period in the history of American ornithology. The major contributions by great ornithologists such as John J. Audubon, William Swainson, John K. Townsend, Charles Bonaparte, and Thomas Nuttall were essentially completed during the first half of the nineteenth century. It was most fortunate that Spencer Fullerton Baird (1823–1887) arrived on the scene during this period and was able to exert a dramatic impact and influence on the exploration of the West. This transition was also crucial in that several capable eastern ornithologists, such as John Cassin, George N. Lawrence, Thomas Brewer, and Robert Ridgway, were able to fill the growing void left by their predecessors and participate under Baird's direction in the monumental work before them.

With a vacancy at the Smithsonian Institution in 1850, Joseph Henry (1797–1878) chose Baird "to take charge of the cabinet and to act as naturalist"

(Rivinus and Youssef 1992). Shortly thereafter, Baird was elevated to assistant secretary, where he emerged unquestionably as the central figure in the field of natural sciences in this country. Through Baird's vision, influence, and organization, the Institution's collections rose in eleven years from six thousand natural history specimens to over one hundred fifty thousand items.

In 1829, the initial bequeathing of funds for this great institution was brought about by the death of James Smithson, an English scientist born of illegitimate but aristocratic origins. Seventeen years later the Smithsonian Institution was officially established. Although Congress had appropriated funding for various natural history collections, monies for the U.S. National Museum, which houses the natural history specimens, were not made available for the new structure until 1879.

Shortly after arriving at the Smithsonian, Baird recognized the possibilities of increasing western military activity similar to Sitgreaves' expedition. He recommended to the army that key men be assigned as naturalists to certain units or be posted at selected army garrisons. Whether they were army surgeons or other officers, as a group they were most energetic and enthusiastic about collecting natural history specimens. Although Baird never had the experience of serving in the army or of traveling west, he had the ability and diplomacy necessary to effect the placement of these men in important locations.

Baird, assisted by John Cassin and George N. Lawrence, compiled and wrote volume 9 of the *Pacific Railroad Reports* (1858). Elliott Coues (1878) stated that this work "exerted an influence perhaps stronger and more widely felt than that of any of its predecessors, Audubon's and Wilson's not excepted, and marked an epoch in the history of American ornithology."

The admiration and respect for Baird's scientific and administrative work is reflected in tribute by two other southwestern naturalists. C. Hart Merriam (1924), the first chief of the biological survey, remarked: "That it should fall to the lot of one man to rise to the highest eminence in science as a result of the value of his own contributions . . . but that [his] personality and influence . . . should prove so stimulating and far-reaching as to create an army of enthusiastic workers . . . [and] move the Congress of the United States to do his bidding, is a thing unprecedented in the history of science."

Henry W. Henshaw, second chief of the biological survey, like Merriam, knew and greatly admired Baird. Of this extraordinary scientist, Henshaw (1920a) wrote: "In his historical summation of American ornithology, Dr. Coues has called the period of activity following the Audubonian era the Bairdian period. The name is well chosen, for America has produced no greater ornithologist."

In a show of affection and to honor Baird, Audubon applied the name Baird's "Bunting" or Sparrow *(Ammodramus bairdii)* to an elusive grassland sparrow that he secured on his trip up the Missouri River. This was the last bird he was to name, and it was added to his seventh and final edition of *Birds of America*. Elliott Coues (1878), recognizing the symbolism of Audubon's thoughtful gesture, wrote: "If a trace of sentiment be permissible in bibliography, I should say that the completion of this splendid series of plates with the name *bairdii* was significant; the glorious Audubonian sun had set indeed, but in the dedicating of the species to 'his young friend SPENCER F. BAIRD' the sceptre was handed to one who was to wield it with a force that no other ornithologist of America has ever exercised." The friendship between Audubon and Baird endured, for the elder naturalist even tutored Baird in the basic principles of painting birds. Baird, no doubt, benefitted from this experience by later publishing serious bird illustrations of his own.

Baird's other major works include *The Birds of North America* (1860), again with Cassin and Lawrence. This was followed, with collaborators Thomas Brewer and Robert Ridgway, by two additional efforts: *A History of North American* [land] *Birds* (1874), and *The Water Birds of North America* (1884).

A colleague of Baird, Thomas Mayo Brewer (1814–1880) was a Boston physician and newspaper writer, and, as an ornithologist, had a serious interest in bird eggs. Brewer's Blackbird and Brewer's Sparrow *(Spizella breweri)* bear his name.

Baird was a mentor for many aspiring men who entered the field of ornithology. In 1864, a young man of fourteen sent Baird a colored drawing of a bird, along with its description, to learn its identity. Thus, Robert Ridgway (1850–1929) embarked on a relationship with the Institution that lasted more than half a century. As curator of birds at the National Museum, he greatly influenced the taxonomy of the fauna in the Southwest. Ridgway's labors and scholarly abilities in the field of science include over 550 titles, with perhaps his foremost contribution being *The Birds of North and Middle America* (1901–1919) written in eight volumes. Following his death, this massive endeavor was completed by his successor, Herbert Friedmann (Ridgway and Friedmann 1941–1950).

Ridgway was a highly productive artist-scientist who portrayed some of the finest examples of North American birds. Unfortunately, his marvelous artistic abilities have been all but forgotten because of his many overshadowing scientific contributions. In addition, he lacked exposure, as most of his significant artwork appeared in limited government publications not readily accessed. Many of the head portraits of birds appearing in *A History of North American Birds* were drawn by Ridgway from specimens collected by naturalists in the Southwest.

Ridgway was commemorated by the naming of the Aztec Thrush *(Ridgwayia pinicola)*, which was placed in its present genus by his assistant Leonhard Stejneger in 1883. He was also a founding member of the American Ornithologists' Union and served as its President from 1898 to 1900. The Ridgway Ornithological Club of Chicago was named in his honor in 1883.

Harry C. Oberholser (1933), a fellow systematist, recognized the harmonious contributions of Baird and Ridgway by stating: "Just as Baird founded systematic ornithology in the New World, just so surely did Ridgway develop and modernize and standardize the technique of the science."

THE WHIPPLE RAILROAD SURVEY, 1853–1854

୬◉ IN MARCH 1853, Congress passed the Pacific Railroad Survey bill. Among the expeditions sent west, several were to explore the southern railroad route possibilities along the 32nd and 35th parallels. Many of the officers and several scientists contributed a series of reports, often scholarly dissertations, which, when published, became a set of great government tomes on the West. These ambitious accounts under the direction of the War Department (1855–1860) consisted of twelve volumes entitled *Reports of Explorations and Surveys to Ascertain the Most Practicable and Economical Route for a Railroad from the Mississippi River to the Pacific Ocean.* This cooperative work was the most exhaustive contemporary assemblage of knowledge on western geography, including human and natural history, to date.

The name and work of Lt. Amiel Weeks Whipple (1816–1863) will long be remembered by those interested in the history of the Southwest. Graduating from the U.S. Military Academy as a topographic engineer, he directed the important railroad survey along the 35th parallel through New Mexico Territory into California during 1853–1854, and later helped determine the boundary between the United States and Mexico. Fort Whipple, situated in central Arizona just south (0°27') of the 35th parallel, the lemon-yellow flowered Whipple Cholla, and the Whipple Yucca, which displays long white flowering stems when in bloom, bear his name.

The expedition to investigate the route along the 35th parallel comprised two units that were to assemble at Albuquerque. While the main party, commanded by Lieutenant Whipple, gathered at Fort Smith before heading west, another contingent, under Lt. Joseph Christmas Ives (1816–1863), began its march from southern Texas. Also among this smaller group was Dr. Caleb Burwell Rowan Kennerly (1829–1862), a young acting assistant surgeon, who was to serve as a naturalist.

Col. Edgar E. Hume (1978), a biographer of ornithologists serving in the Army Medical Corps, recovered Kennerly's diary in which the young doctor wrote of his duties and experiences on that long and difficult journey. Kennerly's concern about the harassing fleas and mosquitoes was well founded, for he contracted malaria almost immediately. This dreadful reoccurring sickness was occasionally noted during his own periodic attacks or when he was treating others in the party who were suffering miserably with the same disease.

Despite the many hardships, Kennerly was in good spirits when he left Fort Inge, Texas on July 27. "This is a task of peculiar pleasure to one far away from those loved ones with whom from childhood he has been accustomed to assemble around the family hearth and enjoying the sweet intercourse which must now be interrupted by so long an interval. . . . Little did I think, when in boyhood's hour, the future seemed so bright, that I should now be a dweller upon the lone prairie, making my bed among the bushes with the wild mustang and wolf."

In addition to collecting and preparing specimens for shipment to the Smithsonian Institution, his duties as contract surgeon were at times unpleasant, as he noted in August: "among the incidents that I have met with . . . to be present while a deserter was [to be] flogged. But, poor miserable wretch, I saved him from undergoing the punishment. He tried an old soldier trick on me, by pretending to have an epileptic fit. . . . [I] exposed it. . . . he was much emaciated and debilitated by disease and sore from previous flogging, and I gave it as my opinion that he would die under the fifty lashes well laid on, so his head was shaved and he was drummed out."

Arriving at Fort Clark, the young doctor experienced another emotional event in mid-September. "This morning things around the camp did not seem so cheerful as usual . . . probably because I was about to witness the burial of one of my patients, the first time that I had ever had that misfortune. He was buried as a soldier, wrapped in his blanket . . . three salutes were fired over the grave. He sleeps now upon the wild prairie with many who have preceded him. The wild wolf may howl over him and the tawny savage yell, but they will disturb him not."

Finally reaching Albuquerque at the end of October, Kennerly wrote of the continuing manifestations of malaria, although he "tried to anticipate [his] chill by taking a large dose of quinine." Disillusioned, he added that it was too late, "for very soon after breakfast it came with such force as to compel me to go to bed, where I was compelled to lie all day, only rising this evening to pay a professional visit."

The parties, having merged in Albuquerque, made preparations to begin their first leg along the 35th parallel. Accompanying Lieutenant Whipple was Baldwin (Heinrich) Möllhausen (1825–1905) who, through a recommendation by the renowned German naturalist Alexander von Humboldt (1769–1858), was accepted by the American government to be a member of this official survey.

By the time Möllhausen was thirty-two years old, he had crossed the Atlantic from Germany to America three times, and taken part in an equal number of westward continental expeditions, with two reaching the Pacific Ocean. Möllhausen's (1969) deep respect for the man who befriended him is evidenced when he wrote: "This fulfil[l]ment of my wishes I by no means owe, however, to my own exertions or merits, but to the untiring kindness of the high-minded man, of whom an American, holding one of the most important public offices of his country, once expressed my own feeling when he said to me, uttering at the same time the general sentiment, 'How sacred to me is every word of Alexander von Humboldt!'"

Through this set of fortuitous circumstances, Möllhausen secured a position from Lieutenant Whipple as naturalist and topographical draughts-man. Many of his fine drawings and illustrations accompany the expedition's official report. It was also in Albuquerque that the two naturalists—Möllhausen and Kennerly—renewed their previous, but brief, acquaintance. Möllhausen had remarked, after their first meeting in Washington, that Kennerly's "frank upright character had inspired me with a warm regard for him, and we had rejoiced together, before the commencement of the journey, at the interesting nature of the employment we were to be engaged upon." From this point on the two naturalists collaborated in every way.

Early in November, with the expedition underway, Möllhausen described the column of soldiers, horses, wagons, and mules as "an imposing appearance, and the van of the long procession stretched out far into the plain, while the last of the muleteers and their beasts were only just leaving the town of Laguna."

West of Laguna, as countless flights of waterfowl migrants were circling, hovering, and screaming about the "spacious lake in the middle of the valley" Möllhausen reflected on the exciting scene. "This is doubtless a common sight enough, yet I can never watch without interest the proceedings of the creatures thus obeying the instinct implanted in them by nature, and making their preparations for these long journeys."

Obviously a man of great comprehension, Möllhausen's reflection on this setting might well serve as one of the great esoteric truisms regarding our natural experiences with wild creatures: "Whoever observes attentively the ever-varying spectacle of busy animal life, and sees, in every movement, in every coincidence, not mere accident, but the wise ordinance of nature, will enter into the pious spirit of those words of Goethe:—'Thus does Nature speak to known, unknown, and mistaken senses, to herself and to us, through a thousand phenomena; and to the attentive observer she is nowhere dead or dumb.'"

Among the great flocks of wintering waterfowl that Möllhausen observed were Snow Geese and, no doubt mixing with them, smaller Ross's Geese *(Chen rossii)*, which were not recognized as a distinct species by Cassin until 1861. It was at the suggestion of Baird that this goose was named in honor of Bernard Rogan Ross (1827–1874), chief trader of Hudson's Bay Company.

On November 19, they reached the "ridge of the Sierra Madre" (Continental Divide), where they began their descent toward Inscription Rock, the same locality that Samuel Woodhouse visited in September 1851. Unlike Woodhouse, neither of the naturalists noted flying swifts about the prominent bluff. This is understandable, for their visit into this relatively high region of over seven thousand feet was late in the season.

Continuing west, they entered an area of fossil trees and crossed the "Rio Colorado Chiquito" (Little Colorado River), which was lined with "luxuriantly" growing cottonwoods, but posed an unseen hazard of "continuous quicksand." Kennerly collected the "beautiful" Black-billed Magpie in this area, which is its extreme southern limit.

After spending Christmas on the south side of the snow-covered San Francisco Peaks, they passed north of Bill Williams Mountain and entered through Partridge Creek into a valley of grama grass they named "Val de

China" (Chino Valley). Proceeding south down the valley, they turned west into "Pueblo Creek," a small stream flowing east from the juniper, oaks, pine, and fir covered "Aztec" (Santa Maria) Mountains. "This was a beautiful little stream," wrote Kennerly (1856), "and we were sorry to leave it, as in this region one seldom sees such cool and limpid water."

On January 22, Kennerly collected an "interesting bird we found inhabiting various points between the Rio Grande and the Great Colorado" that was "very abundant along Pueblo creek." He added that they were in some "respects resembling" House Finches, which they had noted as "very common along the valley of the upper Rio Grande."

Later at the Smithsonian, Spencer Baird, in describing their specimen, consulted John Cassin about naming it *Carpodacus pileatus*. Cassin in a return memorandum wrote: "*Carpodacus pileatus*—bad specific name—greatest bird in the lot—call it *cassini[i]!*" (Palmer 1928a). Baird (1854) followed accordingly: "This species [Cassin's Finch] is named in honor of Mr. John Cassin, of the Academy of Natural Sciences of Philadelphia."

On the same day, Kennerly (1859) wrote they found another "little bird" along the creek and among the "pines of the Aztec Mountains." It proved to be the voluble Bridled Titmouse, and this was the first time it was noted among our United States avifauna. This location represents its extreme northern range.

Entering the source of the "Rio Big Sandy," they followed downstream to the Bill Williams Fork (twelve hundred feet). On February 9, along this segment, they noted Costa's Hummingbird, the first of this species to be added to our fauna, and it, too, was near the northern limit of its range in the lower Colorado River drainage.

Kennerly (1859) wrote, "we found a few flowers that had already expanded beneath the genial rays of the sun, and around these we never failed to find this beautiful bird. At this season they were generally paired, and they were ever flitting around the flowers enjoying their sweets 'with hearts of controversy.'" Apparently, unknown to Kennerly, the behavioral interaction he was witnessing between the hummingbirds during their nesting season was not that of pairing, but rather territorial aggression and sexual promiscuity. Single male hummingbirds generally establish a favorable territory in which they practice polygamy and mate with several females.

Soon they came upon what Dr. John Milton Bigelow (1818–1901), the expedition botanist, described (1856) as by "far the most interesting cactus of the region, and probably the whole world, *Cereus giganteus*." He went on to remark that his observations of the Saguaro "did not accord fully" with the account given by Dr. George Englemann (1809–1884), who named it. Englemann's description was based on the efforts of several men, including Major William Emory who collected seeds, and Drs. Charles C. Parry (1823–1890) and George Thurber (1821–1890), both with the U.S. Mexican Boundary Survey, who collected spines, wood, and specimens complete with flowers.

As they rode down the river, Möllhausen (1969) gave his first impressions of the scene. "If the smaller specimens of the *Cereus giganteus* that we had seen in the morning, excited our astonishment, the feeling was greatly augmented, when, on our further journey, we beheld this stately plant in all its magnificence. . . . They stood symmetrically arranged on the heights and declivities of the mountains, to which they imparted a most peculiar aspect, though certainly not a beautiful one."

On February 13 and 16, while continuing along this stream, they collected the Gila Woodpecker, another new bird that also occurs here at its northern and western limits. Describing it as "gaily variegated," Möllhausen added that they take "up their abode in the old wounds and scars of sickly or damaged" cactus.

The area they were entering defined not only the northern limit of the Saguaro and the huge Sonoran Desert, but the southwestern edge of the merging Mohave Desert. The parameters of the Mohave Desert are marked by another plant indicator, the tree yucca, or Joshua Tree. In the immediate vicinity, groups of these spine-studded tree lilies sometimes consort, side by side, with giant Saguaro.

The two naturalists found the Bill Williams Fork to be the most exciting locality for collecting "new and undescribed species." As they continued their slow march downriver "the beautiful stream sometimes emerged suddenly from the earth a bold rivulet, leaping playfully over its gravelly bed for several miles, and then would as suddenly disappear again beneath the sand. . . . Myriads of ducks and geese were continually frightened from the stream or neighboring lagoons" (Kennerly 1856).

On February 16, roughly twenty-five miles east of the Colorado River, they observed small birds in "large flocks . . . flying and circling around very high. . . . We found them only where the walls of the canons were very high and consisted of almost perpendicular masses of rocks." Instantly recognizing these great aerial performers as swifts, they collected a type specimen of the

White-throated Swift (Baird 1854), a deed which Woodhouse (1853) failed to do at Inscription Rock two years earlier.

On reaching the Colorado River, almost fifty miles south of the 35th parallel, the party turned upstream, passing first among the friendly Chemehuevis. Then, north of the sharply pointed Needles, they assembled in preparation to cross the river as hundreds of Mohave Indians mingled with them to trade and celebrate.

On February 27, crossing began at a point where a small island straddled the five-hundred-yard-wide river and the water was six to twelve feet deep. In finally completing their hazardous ferry by a "much worn India rubber pontoon," Lieutenant Whipple (1856) wrote their "joy in the event was considerably tempered by the accidents that had befalllen" them. Some provisions were lost or ruined and one member of their party and a little Mexican boy came very near drowning "before the exertions of Mr. Möllhausen succeeded in extracting them from beneath the boat."

Resuming their journey west across the Mohave Desert, they passed Soda Lake, the terminus of the Mohave River, then proceeded to its source, the distant San Bernardino Mountains. After reaching the base of the mountains, they diverted through Cajon Pass into the San Gabriel Valley and on to the Pacific Ocean, where their expedition ended on March 23.

Together, Kennerly and Möllhausen collected eighty-eight bird species, several of which have since changed their nomenclature status. Kennerly later participated in the U.S. Mexican Boundary Survey, while Möllhausen, after sailing back to Germany, was invited to return and take part in the exploration of the lower Colorado River under Lieutenant Ives.

THE WILLIAMSON RAILROAD SURVEY, 1853

The year of 1853 was a very active time for the Corps of Topographical Engineers, as several survey parties took to the field. One was a survey to explore all the passes of the southern Sierra Nevada and southern California that had the potential to connect with proposed routes along the 32nd and 35th parallels. Among the several passes investigated in the southern region were the Tejon, New, Cajon, San Gorgonio, and Warner's. The southern Mohave Desert along the Mohave River was also explored and, from as far south as San Diego, routes across the Colorado Desert to the junction of the Gila River with the Colorado

River were examined. Commanding this expedition was Lt. Robert S. Williamson (1824–1882), assisted by Lt. John G. Parke (1827–1900), both topographic engineers who had graduated from the U.S. Military Academy. Williamson had gained extensive experience by leading two surveys in Oregon and northern California. The accompanying naturalist selected was acting assistant surgeon Adolphus Lewis Heermann (1827–1865), a man also with considerable western experience. In his report on the birds of the expedition, Heermann (1859b) listed 220 species and collected 213 specimens, including those from a previous trip to the Pacific coast.

Before joining this expedition, Heermann's first exploits in the west involved a brief trip to the Rocky Mountains in 1845. Then, after returning to Philadelphia, he met William Gambel. Four years later, Heermann (1853b) again traveled west to California, where he stayed for three years. It is from his trip to California that his abilities as a naturalist developed.

After returning to Philadelphia in 1852, the young naturalist was received by John Cassin, who, always anxious to review new specimens, wrote excitedly to Spencer Baird: "Heermann has arrived from California with a collection of about 1200 Bird skins—I have not seen them all, but expect to tomorrow—I have a portion of them, brought in his trunk, amongst which are a Humming bird, *T. Alexandrii,* Bourcier, new to our fauna . . . a lot of nests and eggs" (Dall 1915). The Black-chinned Hummingbird *(Archilochus alexandri)* was first discovered by Dr. Alexandre of Mexico and was jointly named in his honor by two French trochilid collectors, Jules Bourcier and Martial Etienne Mulsant, in 1846. Little is known about Alexandre, who biographer T. S. Palmer described (1928b) as a "man of mystery."

Signor Domiano Floresi, an Italian mining engineer and hummingbird collector, was mistakenly credited as the first to secure the Black-chinned and Costa's Hummingbirds in the "table-lands" and "valleys" of the Mexican Sierra Madre (Baird, Brewer, and Ridgway 1874). Floresi did, however, discover the smallest of the North American trochilids, the Calliope Hummingbird.

While on his trip along the southern California coast, Heermann observed a previously undescribed gull, which he collected and found nesting on Isla Coronado. Overlooked by many preceding naturalists, the seabird was named Heermann's Gull *(Larus heermanni)* by John Cassin (1852) with the following commendation: "I have dedicated this handsome species to my friend Dr. Heermann, as a token of acknowledgment due to his accomplishment as a naturalist, and his great perseverance and success as a scientific traveller." Cassin

Heermann's Gull, Larus heermanni *Cassin, 1852. (Cassin 1856).*

also added the Rufous-crowned Sparrow, another Heermann discovery taken in central California, to the same paper.

While in the Gulf of California, Heermann (1859b) recorded an interaction between two seabirds—kleptoparasitism—a common occurrence between several bird species. Usually the victim is the smaller of the two birds, but his observation was of "a small black [Heermann's ?] gull" incessantly pursuing a much larger Brown Pelican "as the latter plunged into the sea after fish. . . . The pelican emerging . . . to discharge the fluid collected in the gular sac would drop its bill . . . [with] the fish partially protruding . . . the gull would seize upon one and drag it out as his share of the booty. . . . I have never seen the pelican offer the least resistance, or show any anger or impatience at the intrusion or impudence of his little neighbor, who, like a tax gatherer, follows through life, an evil inevitable."

During his stay in Philadelphia in 1852, Heermann spent considerable time working on the academy egg collections, where he compiled a "Catalogue of the Oological Collection in the Academy of Natural Sciences of Philadelphia" (1853a). From this work he is credited with initiating the term "oology" in reference to studying and collecting bird eggs. The practice of emptying eggs through a small hole for the purpose of preserving them and for scientific study began in Europe during the nineteenth century (Brewster 1895).

In June of 1853, Heermann, again restless and looking for more fieldwork, returned to San Francisco as a member of Williamson's Railroad Survey. Situated near this expanding city, Heermann boated to the Farallone Islands, which rise above the sea several miles offshore from the continental shelf. On the islands, surrounded by the nutrient-rich waters of the cold California current, he witnessed the actions of profiteers preying on seabirds. His observations of this disruption to the seabird colonies during the last half of the nineteenth century are historically important. Because of the close proximity of these colonies to the markets of San Francisco, men were able to gather and sell eggs in massive quantities. Among the nesting birds, the Common Murre and the Western Gull were their principal targets.

The vulnerable Common Murre gather on coastal cliffs and island colonies to nest and for protection from their enemy, the Western Gull. During this critical period, gulls rob a limited number of eggs and young, but when combined with the pillages by men, the situation becomes disastrous. Heermann (1859b) described with dismay the typical assault on the helpless murres. "At one o'clock every day, during the egg season, Sundays and Thursdays excepted, (this is to give the birds some little respite,) the egg-hunters meet on the south side of the island. The roll is called. . . . The signal is given, every man starting off at a full run . . . the affrightend murres have scarcely risen from their nests before the gull, with remarkable instinct . . . flying but a few paces ahead of the hunter, alights on the ground, tapping such eggs as the short time will allow before the egger comes up with him. The broken eggs are passed by the men, who remove only those which are sound. The gull then returning to the field of its exploits, procures a plentiful supply of its favorite food."

Heermann noted that among the alcids escaping destruction were those nesting in burrows or rock crevices such as the "sea parrot" or Tufted Puffins and Rhinoceros Auklet. In 1859, Scotsman James Hepburn (1811–1869) also found them breeding on the Farallon Islands, as did Hungarian F. Gruber, who visited the islands three years later (Swarth 1926; Grinnell 1926). These birds then disappeared as breeders for over one hundred years until they re-established in the early 1970s (Ainley and Lewis 1974; DeSante and Ainley 1980). Their recovery has been largely due to the removal of rabbits introduced from England in the late 1800s (Bryant 1888).

Beginning in 1850, the first six years of egg collecting on the Farallones yielded "three to four millions of eggs." Then, fifteen to twenty-five thousand dozen were shipped every year until 1892. By 1896, the "egg picking had fallen to 7645 dozen" (Nordhoff 1874; Taylor 1887; Emerson 1904; Kaeding 1903). At the turn of the century, Leverett M. Loomis (1857–1928), director of the California Academy of Sciences and author of "California Water Birds" (1895, 1896, 1900), became a major participant in enlisting the California Legislature and Theodore Roosevelt to establish protection for the breeding colonies (Bishop 1929). In 1909, the North Farallons were designated a National Wildlife Refuge, but it wasn't until sixty years later that the main islands of the South Farallons were included.

Returning to southern California, Heermann (1859b) collected a pair of Mountain Plover, a bird that shuns water shores in favor of dry plains, unlike other members of Charadriidae. "I first met with this quiet and gentle bird on the plains near the Pueblo Los Angeles, in the month of November, scattered in small flocks industriously gleaning their subsistence over these broad levels."

On the desolate road to Fort Yuma between Carrizo Creek and the Colorado River, Heermann (1859b) described the bounty laid before scavenging Turkey Vultures. "It here finds an ample supply of food from the carcasses of the numerous animals perishing from fatigue or the want of grass and water, and whose whitened bones, strewn over the ground, mark both the road and the hardships of the western pioneer."

Following the completion of his work with the Williamson Railroad Survey in southern California, Heermann was immediately attached to the next expedition heading east across the Colorado River.

THE PARKE RAILROAD SURVEY, 1854

PRIOR TO THE Gadsden Purchase, which was concluded by James Gadsden on December 30, 1853, Jefferson Davis was already arranging for a railroad survey to continue through Mexican territory south of the Gila River. In a letter dated November 18, Davis sent directly to Lt. John Parke the following orders, which read in part: "The President of Mexico has given to this government authority to make surveys within the Mexican territory . . . and you are selected to make such a survey" (Parke 1855).

Parke, a veteran of several years in the West, had assisted Lt. Lorenzo Sitgreaves on the exploration of the Zuni River to its junction with the Colorado and had just completed a portion of the California railroad survey under Lt. Robert Williamson. After receiving Davis's orders on December 20, Parke immediately began preparations for the new route, which was to resume along the 32nd parallel between the Pima villages on the Gila River and Dona Ana on the Rio Grande. On January 24, 1954, the party, consisting of "fifty-six souls" including an escort by Lt. George Stoneman (1822–1894) and twenty-eight men, departed from San Diego. Adolphus Heermann was again selected by Parke to continue as acting assistant surgeon and naturalist.

After crossing the Colorado Desert to Fort Yuma, they continued along the Gila River to the Pima villages, past "castellated" Picacho Peak, and on to the town and presidio "Tuczon" south of "Sierra Santa Catarina." On February 20, Parke and Stoneman rode into the town with some trepidation, but alone, "to allay any fears and correct all misapprehensions on the part of the inhabitants." Their concerns were greatly relieved when the Mexican Army officers addressed them with many "polite attentions and serviceable offices."

For Heermann (1859a), the travels east of the Colorado River into the desert of stately Saguaros provided new opportunities for collecting and observing birds. The Gila Woodpecker, which he had seen on his previous trip to the Colorado River, remained a common bird all the way to "Tuczon." He noted the giant cactus was "frequently filled with holes bored out by this bird. The pith of the plant . . . extracted until a chamber [boot] of suitable size is obtained . . . for the purposes of incubation."

The larger conspicuous woodpeckers of the genus *Colaptes* have long baffled early naturalists, who were generally quick to recognize differences among regional birds. In 1852, the French naturalist Alfred Malherbe first described *C. chrysoides* from a specimen taken in the Cape region of Baja California. However, it was almost a century and a half later before it was separated from the Northern Flicker *(C. auratus)* and recognized as the Gilded Flicker. Heermann apparently recognized the "Red-shafted" and eastern "Yellow-shafted" forms which, incidentally, hybridize on the Great Plains (Short 1965). The flickers in the Southwest also have the potential for local mixing, yet they remain as two distinct species. Heermann in his southwestern travels failed to distinguish these differences. The Northern, or "Red-shafted," Flicker, with salmon-colored wing and tail linings, breeds in the mountains and winters in the adjacent lowlands, while the Gilded Flicker replaces the salmon color with yellow and is resident mainly among the large cactus of the Southwest, including northwestern Mexico and Baja California.

Leaving Tucson and the Sonoran Desert, the party "had no little trouble in effecting a crossing" of the San Pedro River before proceeding onto a broad plain on which is located the "Salt Lake, Playa de los Pimas" (Willcox Playa). With no fresh water available, they continued eastward, causing Lieutenant

Parke (1855) to remark that it was a "perfection of sterility." Passing south of the Dos Cabezas Mountains and through "Puerto del Dado" (Apache Pass), north of the Chiricahua Mountains, they encountered a few friendly Apache Indians before descending into "really rough country" and reaching Apache Springs.

On March 3, the party continued eastward, where they crossed Lt. Col. George Cooke's Mormon battalion trail of 1846. Moving south of the Ojo de Inez, they crossed the Rio Mimbres, stopped at Cooke's "sulphureous" spring, and passed north of the rugged Sierra Florida. This was the area where contention began regarding the boundary dispute between the United States and Mexico, which was ultimately resolved by a much larger acquisition through the Gadsden Purchase. The purchase, at this time, still needed ratification by President Franklin Pierce and the Senate.

It was on this northern extension of the huge Chihuahuan Desert, abounding with Soaptree Yucca, that Heermann collected an Aplomado Falcon. In his report, Heermann (1859a) wrote: "I saw this bird twice on the vast plains of New Mexico, near the United States boundary line, before procuring it; flying over the prairies in search of small birds and mice, and at times hovering, as is the wont of our common sparrow hawk [American Kestrel]." Almost a century after Heermann collected it, this rare raptor disappeared from the United States as a nesting species for reasons unknown, but certainly suggestive of human interference (Ligon 1961; Hector 1980).

The town of Mesilla, located in sight of Dona Ana on the Rio Grande, was finally reached on March 13, and the party was disbanded. In 1855, Lieutenant Parke led another expedition between the Pima villages and the Rio Grande, but Heermann did not accompany it. Heermann (1859a) continued on to San Antonio where he added more birds to his report.

Witmer Stone (1907), associated with the Academy of Natural Science of Philadelphia for over fifty years and president of the AOU from 1920 to 1923, wrote of this naturalist: "Dr. Heermann was one of those pioneers to whom we owe a great deal in the development of our knowledge of western birds. A man who was willing to put up with all kinds of hardship and danger with no other reward than the discovery of new birds or additions to our knowledge of others."

Heermann eventually settled in San Antonio, where he died at a fairly young age. After an inquiry from Stone, Henry Dresser, an English ornithologist visiting Texas, told of Heermann's unfortunate passing. "At the request of his [Heermann's] brother," Dresser had moved into an adjoining bungalow, so

Aplomado Falcon, Falco
femoralis *Temminck, 1822.*
(U.S. War Department
1855–1860)

he could look after the ailing Heermann and also share their common interest
in ornithology. Heermann's deteriorating condition was so disabling that when
riding he strapped "his legs to the saddle." Dresser returned to England, but
continued his correspondence with Heermann's brother and learned of the naturalist's tragic and sudden death. Heermann, suffering from advanced stages of
locomotor ataxia attributed to syphilis, developed neurological problems affect-
ing muscular coordination to the extent that when collecting one day, he evidently stumbled and "his gun going off . . . killed him."

THE POPE RAILROAD SURVEY, 1854–1855

THE CONTINUATION OF the easternmost leg of Parke's survey along
the 32nd parallel was formed under the command of Captain John Pope

(1822–1892), a topographic engineer and graduate of the U.S. Military Academy. An officer with considerable experience in the Southwest, he had received promotions for gallant and meritorious conduct in the war with Mexico. His survey was to traverse the territory between El Paso, on the Rio Grande River, and Preston, on the Red River.

On February 12, 1854, Pope (1855) left El Paso with Dr. W. L. Diffenderfer as surgeon and naturalist. Traveling eastward, they skirted the southern edge of the Organ and Hueco Mountains and continued through Guadalupe Pass to the Pecos. Leaving the banks of the Pecos, they traversed the dry "Llano Estacado, or Staked Plain," crossed the Colorado River, then passed Fort Belknap on the Brazos, before reaching Preston on May 16.

The notebooks written during the expedition reveal little on the avifauna of the region. However, among the specimens of Phasianidae collected there was one new species. Separated at first as a variety of the Greater Prairie-Chicken, the specimens were later recognized as Lesser Prairie-Chicken by Ridgway in 1873.

The great push for transcontinental railroads was postponed by the seceding of the South from the Union in 1861. Although the first railroad to reach the Pacific coast was completed in northern Utah in 1869, the southern routes along the 32nd and 35th parallels were not in full use until 1885.

A FAVORITE PURSUIT, 1851–1857

ARRIVING IN THE lower Rio Grande region in 1851, Dr. Thomas Charlton Henry (1825–1877) began serving first as an acting assistant surgeon and, after two years, he was commissioned as assistant surgeon. During the next six years he was posted at Fort Fillmore on the east bank of the Rio Grande, Fort Webster on the Rio de los Mimbres or "River of Willows," Fort Thorn on the west bank of the Rio Grande, Fort Stanton north of the Sacramento Mountains, and finally Fort Craig, on the west bank of the Rio Grande.

While in the Southwest, Henry "devoted his entire leisure to his favorite pursuit" of observing and collecting birds. However, his excursions in the field were not without excitement. "The first rattlesnake I killed, managed to stick one of his fangs in my finger, before he was quite dead, but with judicious and prompt management and good luck, I escaped with a sore arm for one day only, and the wound healed kindly" (Hume 1978).

Henry followed almost immediately in the footsteps of Col. George McCall of whom he made special reference. During his tenure in the region he compiled two bird lists. Henry's (1855) first list was rather sketchy and consisted of 170 species he "met with, during nearly three years sojourn in New Mexico."

Confusion, identification, and information dispersal regarding taxonomic revisions were considerable for ornithologists such as Henry, who were posted far from other interested scientists or institutions. It was probably through his receipt of the latest journals and correspondence from John Cassin, and perhaps Spencer Baird, that he managed to keep abreast of these changes. Following the end of his sixth year, Henry (1859) compiled a second list of 197 species that was "completely modified." Having his papers edited by Cassin no doubt helped immensely.

While posted in the "Sierra del los Mimbres," Henry observed an "extremely graceful and beautiful shaped jay" that was "very shy and difficult to approach." Once named "Maximilian's Jay," it is now known as the Pinyon Jay because of its close association with the several species of pinyon pine. The name *el pinonero,* given the bird by the Spanish, also notes this close relationship. In Henry's notes, he recognized its discoverer as German savant, Alexander Philipp Maximilian, Prince of Wied-Neuwied (1782–1867), who first secured it while on his great adventure into the trans-Mississippi region in 1833–1834. Maximilian's (1966) great enterprise, almost thirty years after Lewis and Clark, a year earlier than Townsend, and ten years before Audubon, was brought about because of his strong desire to expand scientific knowledge.

Edward M. Kern, while on Lt. John C. Fremont's expedition to California, also collected specimens of this jay in 1846. Col. George McCall, unaware of Maximilian's discovery, mistakenly named it in honor of Cassin in 1851.

In the binomial classification of organisms, certain precepts have been instituted that promote uniformity and prevent confusion. As previously noted in this narrative, species were occasionally reclassified following additional study. A species name, however, given by its original describer, was considered a permanent fixture despite any errors, such as incorrect spelling, color, or geographic locations.

Despite this, unexpected complications arose when Henry (1858) published a "Description of a new Toxostoma from Fort Thorn, New Mexico." In this paper were several printing errors concerning the only two bird species mentioned, a thrasher and a junco. Both birds were named *dorsalis,* quite descriptive of the reddish-brown back of the junco, but totally inappropriate and certainly not distinctive of the uniform brown back of the thrasher. Henry (1859) corrected the mistake the following year by renaming the thrasher *crissalis,* which refers to the rufous crissum of the bird's undertail coverts. Baird also noted the erroneous first printing and mentioned that the page was subsequently canceled and reprinted (Baird, Cassin, and Lawrence 1858). Other discrepancies in the account about breeding and wintering localities were also noted by Florence M. Bailey (1928).

Although the law of priority was more clearly defined in 1889 by the International Zoological Congress, at this time a whole litany of errors ensued, whereby several authors used both names for the thrasher. The transcription error, so easily understood but so difficult to correct, caused taxonomic ornithologist Allan Phillips (1915–1996) to remark: "If the change from *Toxostoma crissale* to '*Toxostoma dorsale*' is condoned, then you have zoology being written by printers' devils instead of by zoologists!" (Phillips, Marshall, and Monson 1964).

A century and a quarter passed before the corrected name given by Henry was finally reinstated, as a rare exception, by the AOU in 1985. The retiring Crissal Thrasher, occurring in the thorny brush thickets of the Southwest, has remained a distinct species, while the ground-dwelling junco described by Henry is considered to be a form of the Dark-eyed Junco.

Another situation involving Henry and the use of eponyms surfaced behind the doors of the Academy of Sciences and the Smithsonian. As in the situation with William Hutton, Cassin voiced strong disapproval of Baird's suggestion of applying an eponym to a nighthawk honoring Lt. John W. Gunnison (1812–1853), who was killed by Indians in Utah. Gunnison's party had secured specimens of the nighthawk, as had Cassin's nominee, Dr. Henry.

In 1855, Cassin wrote to Baird, "I really do not like the idea of calling any bird after Gunnison—he knew nothing about natural history nor never made any exertions in his life at all relating to the matter . . . no more than that you or I should be complimented by military men in a military manner—say with a title of Colonel—Col. Baird or Major Cassin would look very well, but . . . Col. McCaul [McCall] and I held a council on this subject and were unanimous in our conclusion;—as the Col. says Gunnison was undoubtedly a gallant fellow, excellent officer . . . but no naturalist whatever, nor ever demonstrated a fondness for nor encouragement of zoological operations" (Dall 1915). Neither

Gunnison nor Henry was honored in the naming of this bird, because Baird was never able to "define" it as another distinct species, and so it remains today the Common Nighthawk.

HOUSEKEEPING FRONTIER STYLE, 1849–1852

🕉 IN THE SPRING of 1849, Assistant Surgeon Ebenezer Swift (1819–1885) transferred to Brazos Santiago, where a cholera epidemic had started and was spreading to many parts of the southern Texas frontier. After enduring this dreadful experience, he remarked that as "the flowers of the field, men faded away." Then, surviving the full fury of a hurricane that flooded knee deep over the island, sinking or destroying all the vessels, he added gloomily, "the ruin of the town was complete" (Hume 1942).

Three years later, not to be discouraged, Swift transferred, married, and with his bride and her brother, set up housekeeping in a partitioned twenty-four by fifteen foot tent. Fort Chadbourne, his new location, was approximately 185 miles northwest of Austin on Oak Creek, three miles above its junction with the Colorado River. Over the next four years, three sons were born, of which two survived. Life was indeed trying, and warring Comanche Indians nearby only added to their difficulties. Edgar Hume took a passage from the personal memoirs of Gen. David S. Stanley, who recorded Swift's treatment of an ambushed soldier with "fourteen arrows in him, and bristled . . . like a porcupine. Three of these arrows had gone so far through him that the surgeon extracted them by cutting off the feathered part of the arrows and pulling them through the man's body." In two weeks' time, however, the soldier was up and walking.

It was from Fort Chadbourne that Dr. Swift was still able to find time to send to Baird at least fourteen species, including a Red-shouldered Hawk.

A FEMININE APPLICATION

🕉 BEGINNING IN 1849, Assistant Surgeon William Wallace Anderson (1824–1911) served twelve years, with few interruptions, at Forts Polk, Merrill, Terrett, Clark, and Chadbourne, all in Texas and, in addition, Cantonment Burgwin and Fort Union in New Mexico Territory. After six years on the frontier, he went east, married Mary Virginia Childs (1833–1912) and, together, they returned to the West.

It was from Cantonment Burgwin that Anderson collected a lovely new bird species and sent it to Baird, who dedicated it to the discoverer's wife by naming it Virginia's Warbler *(Vermivora virginiae)* in 1860. During the same year, as Anderson's army career was ending in the Southwest, he sent Baird two more boxes of additional specimens. Baird, in receipt of them, replied excitedly: "I found a great many rare birds and other specimens in both, but the Texas box was especially important in having, among other specimens, one of the only species of hawk wanting to our collection, namely Falco femoralis [Aplomado Falcon]. Think of that!" (Hume 1978). This was the second specimen of this rare falcon taken by an early naturalist north of Mexico.

Although married to the daughter of a northern U.S. Army general, Captain Anderson's sympathies remained with his southern heritage, and he switched his allegiance to the Confederate States Army along with his brothers in 1861.

CLOSE ENCOUNTERS, 1854–1856

APPROXIMATELY ONE HUNDRED miles north of Cantonment Burgwin and twenty-three miles east of the Rio Grande at an elevation of eight thousand feet was another isolated post from which Baird received specimens. Fort Massachusetts (Lat. 37°30´) was situated in a region of deep canyons, plunging waterfalls, and rushing streams, which gave Assistant Surgeon DeWitt Clinton Peters (1829–1876) opportunities to observe and collect several birds, including the water ouzel or American Dipper.

In 1854, Peters was attached to the First Dragoons, where he quickly became involved in local Indian skirmishes. One such action "came near costing him his life" when he was following his unit to provide aid to those wounded. From the very start of the ensuing engagement, he had serious problems. First, his frightened mount threw him into "a thrifty bed of prickly pears, the thorns of which did not, in the least, save [him] from being hurt." After regaining his saddle, he was joined by a soldier whose horse had broken down. As they "advanced together . . . by some large sand hills, behind which several Indians sought refuge . . . they made a dash at us and commenced firing their arrows in fine style. My horse now became unmanageable, and by some unaccountable impulse made directly for the Indians . . . [it] seemed determined to bring me into uncomfortably close quarters with a young warrior, who constantly turned and saluted me with arrows. As the situation was getting decidedly unpleasant, I raised myself in the saddle, and sent a ball from my revolver through the body of the Indian" (Hume 1978). With moments like these, collecting birds, no doubt, was a tame event for the young doctor.

While in northern New Mexico Territory, Peters, a great admirer of Kit Carson, gained a lasting friendship with the famous frontiersman and became his authorized biographer. In 1856, still at Fort Massachusetts, he resigned, but he re-entered the army four years later. While serving at Fort Davis in Texas, he was taken prisoner by the Confederate States Army.

Following the Mexican War, Assistant Surgeons William Shakespeare King (1810–1895) and John Fox Hammond (1820–1886), who were posted at various installations in Texas, New Mexico, and Southern California, also contributed bird specimens to the Smithsonian Institution.

4 THE NEW BOUNDARY AND ON TO THE PACIFIC

"Well do I recollect the ride from Sonoyta to Fort Yuma [about 120 miles
NW] and back, in the middle of August, 1855. It was the most dreary and
tiresome I have ever experienced. Imagination cannot picture a more
dreary, sterile country, and we named it the 'Mal Pais' [badlands]. . . .
The eye may watch in vain for the flight of a bird; to add to all is the
knowledge that there is not one drop of water to be depended upon."
  —LT. NATHANIEL MICHLER (Emory 1857)

"[B]efore the burning heat has withered the freshness and beauty of
the early vegetation, this valley [Mojave on the Colorado River] . . .
appears in the most attractive aspect. It may be that the eye, weary of
the monotonous sterility of the country below, is disposed to exaggerate
its charms, but as we first saw it, clothed in spring attire, and bathed in
all the splendor of a brilliant morning's sunlight, the scene was so lovely
that there was a universal expression of admiration and delight."
  —LT. JOSEPH CHRISTMAS IVES (1861)

IN 1849, A year after the Treaty of Guadalupe Hidalgo, the boundary
survey began between the United States and Mexico. Although major portions
of the boundary line had been agreed upon, namely the topographic separation by the Rio Grande between Texas and Mexico, and the California border
between Yuma and San Diego, a sizeable area immediately north of El Paso and
west of the Rio Grande remained in question.

The situation had become politically charged in the United States and
was compounded by misinterpretations of boundary lines and mismanage-

ment, when the Gadsden Treaty, a vehicle for solution, resolved the issue in 1853. Through this action, the United States purchased from Mexico certain lands south of the Gila River and between the two great south-flowing regional rivers: the Colorado and the Rio Grande. After almost seven long years, the work of the Boundary Commission was completed in the fall of 1855.

THE UNITED STATES–MEXICAN BOUNDARY SURVEY, 1849–1856

MAJOR WILLIAM H. Emory, who had worked in various capacities on the Mexican Boundary Survey from its inception, concluded the difficult task a little over a year after his appointment as the new boundary commissioner in 1854. The expertise of the Corps of Topographical Engineers, with experiences of exploration and the railroad surveys, proved equal to the challenge.

Major Emory, a much respected officer, known by his classmates at West Point as "Bold Emory," was twice brevetted in the Southwest for gallant and meritorious conduct in Mexico and California and, in addition, for meritorious and distinguished services as boundary commissioner. Two conspicuous plants occurring along the border honor him: Emory Baccharis, a unisexual flowering shrub in which the pistillate plants produce great amounts of "fuzz" or "cotton," and the evergreen Emory Oak. In the Big Bend region, Emory Peak (7,835 feet), the highest point in the Chisos Mountains, was also named in tribute to this outstanding engineer-soldier.

The *Report on the United States and Mexican Boundary Survey* (Emory 1857, 1859) is a quarto of three volumes which, in addition to describing the work of the surveyors, serves as an extension of the massive *Pacific Railroad Reports* by the U.S. War Department (1855–1860). Spencer Baird wrote "Birds of the Boundary," which was included in volume 2, but it contributes little more than a listing of 225 border species that are also referenced to the 738 enumerated in volume 9 of the earlier report. Baird also illustrated thirty-seven species depicted on twenty-five chromo-lithograph plates. *The Birds of North America* (1860), written by Baird, Cassin, and Lawrence, revised both reports.

In the late 1840s, Spencer Baird, while at Dickinson College, instructed John Henry Clark (ca. 1830–1885) and later secured a job for him to participate in the exploration of the Southwest. In 1851, Clark joined the Mexican Boundary Survey, first serving under topographic engineer Col. James D. Graham (1799–1865) and later becoming principal assistant astronomer. Following a dispute between Graham and Boundary Commissioner John R. Bartlett (1805–1886), in which both were ultimately removed, Clark became principal assistant astronomer under Major Emory. Judging from the locations of where Clark collected specimens, he traveled throughout southern and western Texas, including the adjacent states in Mexico, but only as far west as the Copper Mines (Fort Webster) in New Mexico Territory.

In 1851, Clark, while in El Paso, was the first to add to our fauna a migrating Broad-tailed Hummingbird. Although he discovered it in Mexico, Baird admitted some uncertainty about the bird, stating: "I am indebted to Mr. John Gould, who identified it when examining the specimens of Humming Birds preserved in the Smithsonian Institution" (Baird, Cassin, and Lawrence 1858).

Of related interest, in 1887, Gould, the well-known English artist considered to be a leading authority on Trochilidae, observed his first living hummingbird while on his visit to Philadelphia. Of this momentous occasion, he remarked "my earnest day thoughts and not infrequent nightdreams of thirty years were realized by the sight of a Humming Bird [Ruby-throated] . . . my wish was gratified by the sight of a single male in the celebrated Bartram's Gardens" (Sauer 1982).

Camp life on the boundary survey was generally quite busy with many chores to complete. Still, idle time was not the exception for the "muleteers" who sometimes engaged in dispatching "Cow Birds" [Brown-headed]. Clark described the situation. "Very tame. . . . Sometimes following the herd of mules, making themselves very familiar with them by perching in numbers sometimes on their backs. So tame are they that the herders often indulged in the cruel sport of killing them with their whips" (Baird 1859a).

Clark also traveled in northern Chihuahua, Mexico, where he secured a grebe to which George N. Lawrence applied the species name *clarkii* in honor of its collector in 1858. In the same year, Lawrence named a very similar grebe *occidentalis,* based on a specimen taken in central California by Lt. William P. Trowbridge (1828–1892), a graduate of the U.S. Military Academy serving with the Coast Survey. Elliott Coues (1874a) and Henry Henshaw (1881), however, suggested that they were not two species, but only two color phases. From 1886 to 1985, the grebes were classified as one species and, during the same period, were also placed in a new genus, together described as Western Grebe *(Aechmophorus occidentalis).* Then, in 1985, the AOU, after considering the morphologic differences of the two sympatric forms and that they were reproductively isolated and functioning biologically as "separate species," reinstated them as

Clark's and Western Grebes along with their original species names (Ratti 1979; Nuechterlein 1981).

Clark and Baird frequently corresponded, and in a letter to Baird, Clark wrote of some of the hardships and desolation encountered during his travels. "We reached El Paso on the 25th of June, having made the trip in 45 days; of all the barren waterless regions on the face of the earth I want to see no worse than I experienced on this route. There are stretches of from 50 to 100 miles without living water, without grass, and without wood enough to boil a pot of coffee" (Dall 1915).

Baird responded by recognizing Clark's trying circumstances with words of encouragement and praise. "One of these days, when the results of the expedition are published, people will be astonished to find how much one man can do under difficulties." Referring to their earlier college association, Baird added a fatherly quip: it "all depends on training, don't it."

Although not officially participating in the surveying aspect on the border, Lt. Darius Nash Couch (1822–1897) was an officer who contributed significantly to the avifauna of the region. After graduating from the U.S. Military Academy, he served in the Mexican War and was brevetted for gallant and meritorious conduct in the battle of Buena Vista. In 1853, he became so interested in the natural history of the region that he took a leave of absence to explore more of northeastern Mexico. Baird (1859a) encouraged the young officer and he added Couch's collections to the "Birds of the Boundary."

While in Nuevo León, Mexico, Couch (1854) collected a beautiful oriole that he believed to be a new species and named it Scott's Oriole *(Icterus scottii).* "I have named this handsome bird as a slight token of my high regard for Major General Winfield Scott, Commander and Chief of the U.S. Army." A very successful and highly decorated officer during the Mexican War, General Scott was quite ceremonious in dress, which gained him the reputation of "Fuss and Feathers."

Unknown to Couch, Charles Bonaparte (1837) had named the same bird *Icterus parisorum,* troubling ornithologists to the present about the origin of this eponym. Often quite vague in his brief descriptions, Bonaparte sometimes failed to include significant background data, as in this instance. In the preface of a paper, which included several new birds, he acknowledged "the kindness of the Messrs. Paris" in allowing him "to examine their small collection" from Mexico. The full names of these men has never been determined but they were probably living in Paris and obtained specimens from abroad as part of their business (Choate 1985). Continuing, Bonaparte wrote:

> *I have much pleasure in naming this bird after the brothers Paris, who, notwithstanding the arduous nature of their professional engagements in Mexico, allowed no opportunity of furthering the interests of science to pass unimproved. I quite agree with the opinion, that in a country whose commercial transactions are so extensive as they are in this, the captain of a trading-vessel bringing home "a 'curious bird,' which may prove to be new, has no claim to have his name immortalized;" but the same rule I would not apply to the Roman state, where a person crossing the sea is a rare occurrence.*

Although the scientific name applied by Couch was invalidated by Bonaparte's description seventeen years earlier, General Scott is still honored by the common name.

Showing sensitivity unusual for collectors of the period, Couch described his experience of securing an Audubon's Oriole, formerly named the "Black-headed Oriole."

> *It was the day after a severe* norther, *and the whole feathered kingdom was in motion. My guide soon called my attention to two* calandrias, *as these birds are called by the Mexicans, which were quietly but actively seeking their breakfast. The male having been brought down by my gun, the female flew to a neighboring tree, apparently not having observed his fall; soon, however, she became aware of her loss, and endeavored to recall him to her side with a simple* pout pou-it, *uttered in a strain of such exquisite sadness, that I could scarcely believe such notes to be produced by a bird, and so greatly did they excite my sympathy, that I felt almost resolved to desist from making further collections in natural history, which was one of the principal objects of my journey. (Cassin 1856)*

Although Couch was not the first to collect Audubon's Oriole, or had anything to do with its naming, the origin of its present common name is of some interest as a Texas borderland bird. Jacob P. Giraud, Jr., an amateur bird collector, produced "A Description of Sixteen New Species of North American Birds" (1841), in which he named this oriole *Icterus audubonii*. Living in New

York, he knew John Bell and John J. Audubon, and gained the acquaintance of Spencer Baird, who said his collection was "the finest I have seen" (Rivinus and Youssef 1992). However, questions arose as to whether Giraud's "sixteen new species" had already been described and where they had been collected.

One of the early skeptics of Giraud's collection was English ornithologist Philip L. Sclater (1855), followed by Elliott Coues (1878), who wrote with some uncertainty, "Doubt is usually entertained that these birds were taken in Texas; but the author stoutly so maintained to the day of his death, and recent discoveries along our southwest border render it more probable than it formerly seemed. Most of the species have been identified with earlier named ones."

Finally, Harry C. Oberholser, after some review, wrote in *The Bird Life of Texas* (1974) that "it is now fairly well established that most, if not all, of Giraud's specimens had been obtained in Mexico." Ultimately René Primevère Lesson's earlier scientific name of *I. graduacauda* of 1839 was retained and Audubon is still honored by the common name.

The "White-necked Crow," or Chihuahuan Raven, was also discovered by Couch while in Tamaulipas, a Mexican state adjacent to Texas. After Couch described the raven in 1854, he heard of another "White-necked Crow" that was found in Texas. In a note to Baird, he wrote, "I am glad that [Capt. John] Pope has found my crow in the U.S. but as my specimen was taken only 20 miles from the Rio Grande I thought it very strange if his majesty did not occasionally come over to enjoy democratic freedom" (Dall 1915).

In 1858, Baird named Couch's Kingbird *(Tyrannus couchii)* in honor of this young officer who discovered it in the Mexican State of Nuevo León, Mexico. It was almost a quarter of a century later before it was observed north of the border (Sennett 1878), and it was among several birds painted by Baird (1859a) in "Birds of the Boundary."

It was through the efforts of Couch that some items from the collection of Swiss naturalist Jean Louis Berlandier (1805–1851) were included in the survey. In 1826, Berlandier, primarily a botanist, arrived in Mexico and, deciding to remain, became a physician. On a trip to Mexico City, he drowned while crossing Rio San Fernando on horseback. Couch, after a series of negotiations with his widow, wrote Baird that he was able to purchase the large collection for the Smithsonian Institution at his offered price ($500) adding: "The payment will cramp me a little, and delay my departure into the interior" (Dall 1915). Of the several birds included in the collection, two species were added to the boundary survey list from Matamoras, Mexico. They were the graceful Reddish Egret and the eggs of the tiny Least Grebe.

Another officer, Capt. Stewart Van Vliet (1815–1901), collected nine species while at Brazos, Texas, including the Black Skimmer, which also breeds into South America. When viewing these birds in Argentina in 1832, Charles Darwin ([1839–1843] 1962) referred to them as "Scissor-beaks" as "they ploughed . . . their course" with mouth agape and large lower mandible submerged, "skimming," or slicing, the water surface in anticipation of fish.

Baird also added the Thick-billed Parrot to the *Mexican Boundary Survey* (Baird 1859a) and *Pacific Railroad Reports* (Baird, Cassin, and Lawrence 1858) by reason of a specimen housed "in the collection of the Philadelphia Academy of Natural Sciences, labeled Rio Grande, Texas, J. W. Audubon" (Sclater 1857c).

In 1837, the year after Texas became a republic, a young Yale graduate, somewhat eccentric in manners, arrived amid all the turmoil that accompanies a frontier region immersed in friction and change. First as a teacher, then as a surveyor, always a botanist, Charles Wright (1811–1885) remained in Texas for the next fifteen years, except for a visit to Harvard at the invitation of botanist Asa Gray (1810–1888). During this visit, Gray, desirous to obtain plant materials out west, arranged for Wright to accompany the Mexican Boundary Survey. Unfortunately for Wright, however, no positive assurances for either transportation or rations were provided. Nevertheless, Wright, eager to collect, accepted the challenge. It was certainly to his credit and sheer determination that he was able to endure the hardships of this frontier adventure.

On May 31, 1849, after considerable difficulties reaching San Antonio via Galveston and Austin, Wright wrote Gray as he was about to leave for El Paso. "[I] put my baggage on the waggon without the least assurance of subsistence and I have been obliged to muster up all the Yankee confidence natural to me (which *is* and always *was* but *little*). . . . I shall start after the train this evening and I shall try to get something to eat out of somebody. . . . There is still some cholera here. . . . I must start this evening and walk 15 miles to overtake my waggon—not a very pleasant evening to walk either in prospect or in execution" (Geiser 1937).

On June 2, disgusted and discouraged, he added: "I have money in my pocket but it does me no good I can buy nothing with it I sit uninvited and see others eating and it is a severe trial to my feeling to *thrust* myself among them The men have their rations and often none to spare . . . I am fully resolved that this season will close my botanical travels on horseback or on foot if I can not operate to better advantage I'll give it up and turn my attention to something else."

After finally arranging to mess with the transportation train, Wright's apparent worry for his needs was still manifested by his suspicious nature and unsettling comment about the possible actions of the quartermaster officer in charge of the caravan. "Whenever Capt. [Samuel G.] French gets in an ill humor he begins to grumble about the weakness of his teams and the transportation of botanists' tricks." In spite of all his difficulties, Wright arrived in El Paso 104 days later, walking the entire distance of 673 miles.

Gray, grateful for Wright's collection, wrote some thirty years later: "You cannot over-estimate the services which Charles Wright has rendered to Botany. He has been not only a capital and indefatigable explorer and collector, but also an acute observer." Many plants were named for him by such renowned botanists as Asa Gray, John Torrey (1796–1873), and George Engelmann (1809–1894), who named one small cactus Wright's Fishhook.

Wright's expectations of joining the Mexican Boundary Survey did not materialize, so he again fell back to teaching on the frontier. In the spring of 1851, however, he managed to join Col. James D. Graham's party as a surveyor and, in addition, he expanded his collecting to include avifauna.

Wright's acceptance and recognition broadened when, near El Paso, he secured a bird later named in his honor. It did, however, add another taxonomic problem to an already confusing group of small gray flycatchers with tinges of brown, olive, or yellow that have two pale wing bars and light eye rings. Baird, in addressing the genus *Empidonax* to which it belonged, wrote: "There is no species of North American birds more difficult to distinguish than the small flycatchers, the characters, though constant, being very slight and almost inappreciable, except for the acute observer" (Baird, Cassin, and Lawrence 1858).

Baird identified the empid, but in doing so, he stated with some reservation, "For the present, therefore, I retain the name *obscurus*, but should this prove distinct, shall claim that of *E. wrightii*, the discoverer, by which I had provisionally designated it." William Brewster (1889), in reviewing this complex genus three decades later, honored Baird's statement of priority. "Obviously it [*obscurus*] can serve no longer for Baird's bird which must stand hereafter as *E. wrightii*."

Fifty years elapsed before Allan R. Phillips (1939) again reviewed the genus. His research revealed that Wright's specimens actually belonged to *E. gri-seus,* which he promptly voided and, because of priority, applied to that group *E. wrightii,* the Gray Flycatcher. In the confusing process of re-designation, Phillips renamed the remaining assemblage of closely allied specimens *E. oberhol-seri,* the Dusky Flycatcher. Phillips, in naming this last species for taxonomist Harry Church Oberholser, wrote of his "unfailing assistance" in unraveling the complexities of "this troublesome genus."

Several of the naturalists who worked on the Mexican Boundary Survey had prior western experience, including Dr. Caleb B. R. Kennerly, who was not only a student of Spencer Baird, but a veteran serving with Baldwin Möllhausen on Lt. Amiel Whipple's survey along the 35th parallel. Although listed as a surgeon, Kennerly participated in the Mexican Boundary Survey as a naturalist along the Texas border west to "Los Nogales" in New Mexico Territory.

In addition to collecting, Kennerly was a keen observer, as evidenced by his remarks about the hunting techniques of a Ferruginous Hawk at "a prairie dog town" near Fort Davis, Texas. His notes tell how the hawk was "intently watching at a hole of one of those animals. While in this position, [the raptor] was observed to strike at the prairie dog [Black-tailed] with its claw when its head was protruded" (Baird 1859a).

Kennerly's notes on the "Massena Partridge" or Montezuma Quail reveal some of the reasons why these birds have received the unwarranted aberration of "fool quail" based on human misconception. Their natural instinct when alarmed is to freeze and rely on their cryptic coloration, which in the following instance worked to their detriment. "When hunted it hides itself very closely in the grass, and often I have known the Mexican soldiers in Sonora to kill them with their lances by striking them either while on the ground or just as they rise. Some of these men are very expert in this business, and will kill many in the course of a day's travel" (Baird 1859a).

After spending a third of his young life in the American West, Kennerly died in 1862 during a shipwreck off the west coast of Mexico en route home to be married. Elliott Coues (1878) remarked that Kennerly's "death, under very deplorable circumstances, left a gap in the ranks of western explorers."

Several survey parties were in the field with different sectors assigned to them. In addition to a stretch from Fort Duncan to Laredo on the Rio Grande, Lt. Nathaniel Michler's (1827–1881) major surveying segment was east from Fort Yuma to the 111th meridian and included some of the most arid country of the entire border. During this western portion of the survey, Michler's group was out of touch with Emory much of the time, enduring hunger, thirst, desertions, thunderstorms, and even death to one member of the party.

This desolate area of the Sonoran Desert was given the reputation of *tierra incógnita,* or "unknown lands," by the Spanish soldier Captain Juan Mateo Manje, the usual escort of Father Eusebio Kino at the turn of the seventeenth century. Soldiers, explorers, prospectors, and missionaries journeyed through the region on the *Camino del Diablo,* or "Devil's Highway," between Mexico and California. Known water sources were scarce, resulting in more than four hundred unfortunate wayfarers perishing of thirst in the 1850s.

It is also a land of contrasts, for in early spring, depending on elevation and rainfall, a world-class display of annual flowering plants occurs in this desert. Two showy drought-deciduous trees also turn into a blaze of color as they flower in spontaneous bloom. First, the Blue Paloverde, tall and wildly irregular, unfurl a brilliant yellow ribbon as they wind through low desert washes. They are followed by the smaller Foothill Paloverde, more uniform and widespread, clothing the rocky slopes and hills with a softer, but stunning pale yellow glow. Although the flowers of these trees are quite similar, the brilliance of the latter species is much less vivid—having one white petal on an otherwise all yellow flower.

Michler's account of surveying this isolated strip in temperatures up to 120°F. exemplifies their difficulties. "Even in winter the sun is so hot, and the direct as well as reflected light upon the sand-plains so dazzling, that, excepting a couple of hours after daybreak and an hour before sunset, it is only possible to see objects through the best instrumental telescopes in the most distorted shapes—a thin white pole appearing as a tall column of the whitest fleece" (Emory 1857). To reduce distortion under these conditions, most surveyors in later years employed night lights as targets when sighting long distances.

Arriving from San Diego, Michler established his starting point "on the Colorado River twenty English miles below its junction with the Gila," and then proceeded on an easterly azimuth to a ridge on the "Tinajas Altas." Michler noted the water pools for which these mountains are named and a "small, delicate humming-bird [Costa's ?]" in the area. Unable to follow their desired line further east because of limited water, they returned to the Gila River, where they traveled east and then south to the intersecting parallel of 31°20' with the 111th meridian. From this point in the "Sierra de los Pajaritos" (the Mountains of Little Birds), the party resumed surveying in a northwesterly direction, setting monuments as they traversed along the following route: passing the southern end of the "Sierra Babuquivari," across the Sonoyta Valley to within "a few feet south of the springs at Quitobaquito," and finally south of the "Tinajas del Tule" before completing the line back to the "Tinajas Altas." Immediately adjacent to the route on the Mexican side of the boundary is the spectacular volcanic field of the Sierra Pinacate. Cerro Pinacate (4,060 feet) dominates an area of ancient lava flows, cinder cones, and giant hydro volcanic maar craters, which result from superheated steam blasts. After beginning at Fort Yuma on December 9, 1854, the survey of this diagonal Arizona/Sonora boundary segment (about 235 miles) with its overriding circumstances was not completed until August 25, 1855.

During the fieldwork, Michler's concern about the number of rattle-snakes they encountered reveals some of his rather interesting speculations regarding their nocturnal visitations and their avian enemies.

> *When you lie down on your blankets, stretched on the ground, you know not what strange bedfellow you may have. . . . My servant insisted upon encircling my bed with a reata of horse-hair to protect me from their intrusions. Snakes are said to have a perfect repugnance to being pricked by the extremities of the hair. The paisano, or chap-paral cock [Greater Roadrunner], surrounds his antagonist, while asleep, with a chain of cactus thorns; when the preparations are all made the bird flutters over the head of the snake to arouse it to action; the latter, in its vain efforts to escape, is irritated to such a degree, by running against the barrier encompassing it, that it ends its existence by burying its fangs in its own body.*

Lieutenant Michler also noted that accompanying the Mexican commission was a "Senor Augustin Diaz [1829–1893], 2d engineer," who assisted in the verification of the survey boundary line. He was an engineering officer during the war with the United States, served on the Mexican Boundary Survey, and later became professor of engineering at the Mexican Military College. He was a founder of the Mexican Geographical and Exploring Commission, serving as its director beginning in 1866. In the same year, Robert Ridgway named the Mexican Duck *Anas diazi* in his honor, but it was later merged as a single species with the Mallard *(A. platyrhynchos)* (Aldrich and Baer 1970; Hubbard 1977).

Among several civilians attached to the Mexican Boundary Survey was Arthur Schott (1814–1875), an able Austrian collector, geologist, surveyor, and artist, who served as assistant to Lieutenant Michler. Among several plants common to the border region named for Schott are the Pygmy Cedar *(Peucephyllum schottii)*, Mountain Yucca *(Yucca schottii)*, and Schott Calliandra.

The notes regarding his bird collecting only specified locations of where they were taken, but they reveal that his work was extensive, and it included much of the border from California to Texas. While along the Rio Grande, Schott appears to be the first naturalist to add the Red-billed Pigeon to our fauna north of Mexico (Baird 1859a).

Schott also related to Emory (1857) a harrowing experience that occurred between their party, including a contingent of fifteen soldiers, and a band of forty "Kioways and Comanches who had recently returned from a foray into Mexico with nearly one thousand animals." As their party approached Comanche Springs (later Fort Stockton), which the Indians occupied, Schott considered it an "inevitable fight, or die with thirst" situation. Without faltering, the soldiers pushed with determination to the spring, where they "corraled" their wagons while one of their party climbed a nearby hill, pretending to look for reinforcements. The bluff apparently worked, as Mucho Toro, the Indian leader, soon visited their camp peacefully, displaying an "immense silver cross" draped around his neck, declaring it was a gift from the Bishop of Durango. After sketching a portrait of him, Schott quipped: "He had, no doubt, robbed some church of it." Following their deception and brief parley with the Indians, the party escaped just before the main body of over four hundred warriors arrived.

STEAMBOATING THE COLORADO, 1857–1858

THE POTENTIAL OF a war with the Mormons brought about the need to explore the lower Colorado River as a possible military supply route to the interior southwestern army garrisons. In 1852, a civilian company began supplying Fort Yuma via the Gulf of California by a steamboat, the *Uncle Sam*. This successful venture encouraged the possibilities of this method of transportation. The account of this logistically difficult enterprise appears in the interesting *Report upon the Colorado River of the West* (1861) by Lt. Joseph C. Ives.

In 1857, Ives, who previously assisted Lt. Amiel Whipple on the railroad survey along the 35th parallel, was instructed to "ascertain how far the [Colorado] river was navigable for steamboats." Upon receiving his orders, Ives sent a transmittal to his German friend Baldwin Möllhausen (1969), who had accompanied him, along with Dr. Caleb Kennerly, on that same memorable railroad survey. "I am desired by the Secretary at War to communicate to you that you have been appointed as assistant to an expedition to proceed under my command to the exploration and survey of the Colorado River." The favorable regard Ives had for this talented artist-naturalist was indeed warm, for he was inviting Möllhausen, at the age of thirty, to again cross the Atlantic for a third trip to America.

To accomplish the river cruise, Ives (1861) placed with a Philadelphia company an order consisting of "an iron steamer, fifty feet long, to be built in sections, and the parts to be so arranged that they could be transported by railroad . . . *via* the Isthmus of Panama" to California.

By mid-October, the party had assembled in San Francisco and divided into three detachments: Dr. John S. Newberry, a geologist in charge of the "natural history department," traveled to San Diego to obtain mules before proceeding to Fort Yuma; Möllhausen accompanied another group for the same purpose to Fort Tejon, where, incidentally, he met the Hungarian naturalist John Xántus; and Ives, with the unassembled steamboat lashed aboard the *Monterey,* sailed southward around Cape St. Lucas, entered the Gulf of California, and then turned northward to the mouth of the Colorado River.

Leaving San Francisco on November 1, the *Monterey* and her company encountered brisk breezes, "dead calms, burning tropical days," glassy swells, and gale-like winds for thirty days before anchoring at "Robinson's Landing," 150 river miles south of Fort Yuma in the Colorado River delta. Here, resting on pilings extending four feet above the surrounding tidal mud flats, stood a small deserted house. Near this anchorage, close to where the *Explorer* was to be assembled, Ives observed that "Innumerable flocks of pelicans, curlews, plovers, and ducks of different varieties, were scattered over the flats."

In the succeeding days, the men laboriously dug a large pit and erected the cribbage necessary for assembling the steamboat in preparation for it to be floated off during the next full moon tide. The careful selection of the pit and the use of the tide was essential for the success of the operation.

Their anticipation of the rising sea was quite unlike that of the first fur trappers who penetrated this region a little more than three decades earlier in 1828. James O. Pattie (1984), accompanied by a small group of men, wrote about the sudden unexpected occurrence of a tidal surge that, in some seasons, forms a twenty-foot bore at the confluence of the "Red [Colorado] river" and the sea. "We made our canoes fast to some small trees, and all lay down to sleep, excepting my father, who took the first watch. He soon aroused us, and called on us all

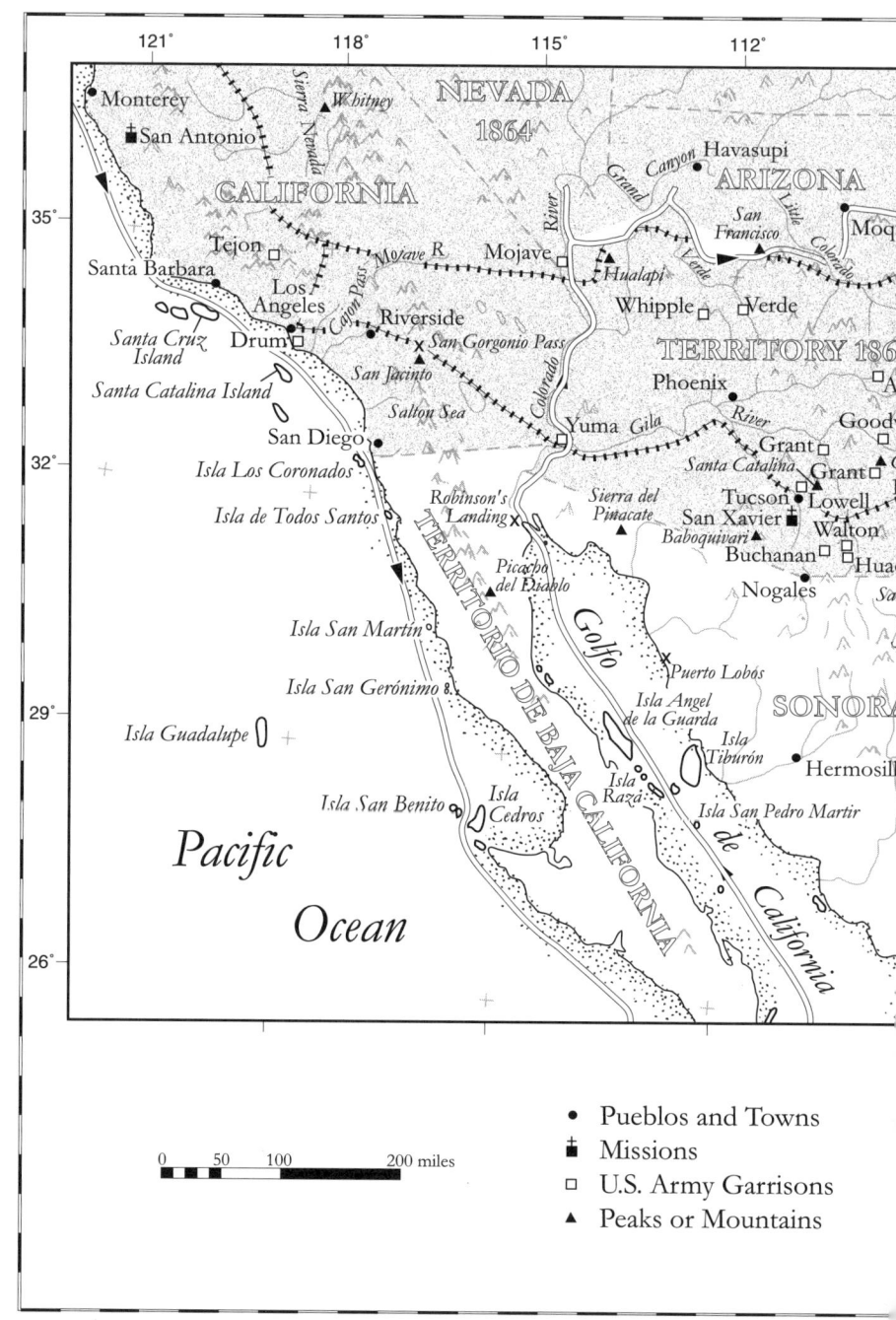

Monterey
San Antonio
Sierra Nevada
Whitney
NEVADA 1864
CALIFORNIA
Grand Canyon
Havasupi
ARIZONA
San Francisco
Little Colorado
Moqui
35°
Tejon
Mojave R.
Mojave
Hualapi
Santa Barbara
Los Angeles
Canyon Pass
Riverside
Whipple
Verde
Verde River
Santa Cruz Island
Drum
San Gorgonio Pass
San Jacinto
Salton Sea
Colorado
TERRITORY 1863
Ap
Santa Catalina Island
Phoenix
Gila River
Goodw
Yuma
32°
San Diego
Grant
Isla Los Coronados
Santa Catalina
Grant
G
Isla de Todos Santos
Robinson's Landing
Sierra del Pinacate
Tucson
Lowell
B
San Xavier
Walton
Baboquivari
Buchanan
Huac
Picacho del Diablo
Nogales
San
TERRITORIO DE BAJA CALIFORNIA
Golfo
Isla San Martín
Puerto Lobos
29°
Isla San Gerónimo
Isla Angel de la Guarda
SONORA
Isla Guadalupe
Isla Tiburón
Isla Raza
Hermosill
Isla San Benito
Isla Cedros
Isla San Pedro Martir
Pacific
Ocean
de California
26°

0   50   100   200 miles

• Pueblos and Towns
✚ Missions
▫ U.S. Army Garrisons
▲ Peaks or Mountains

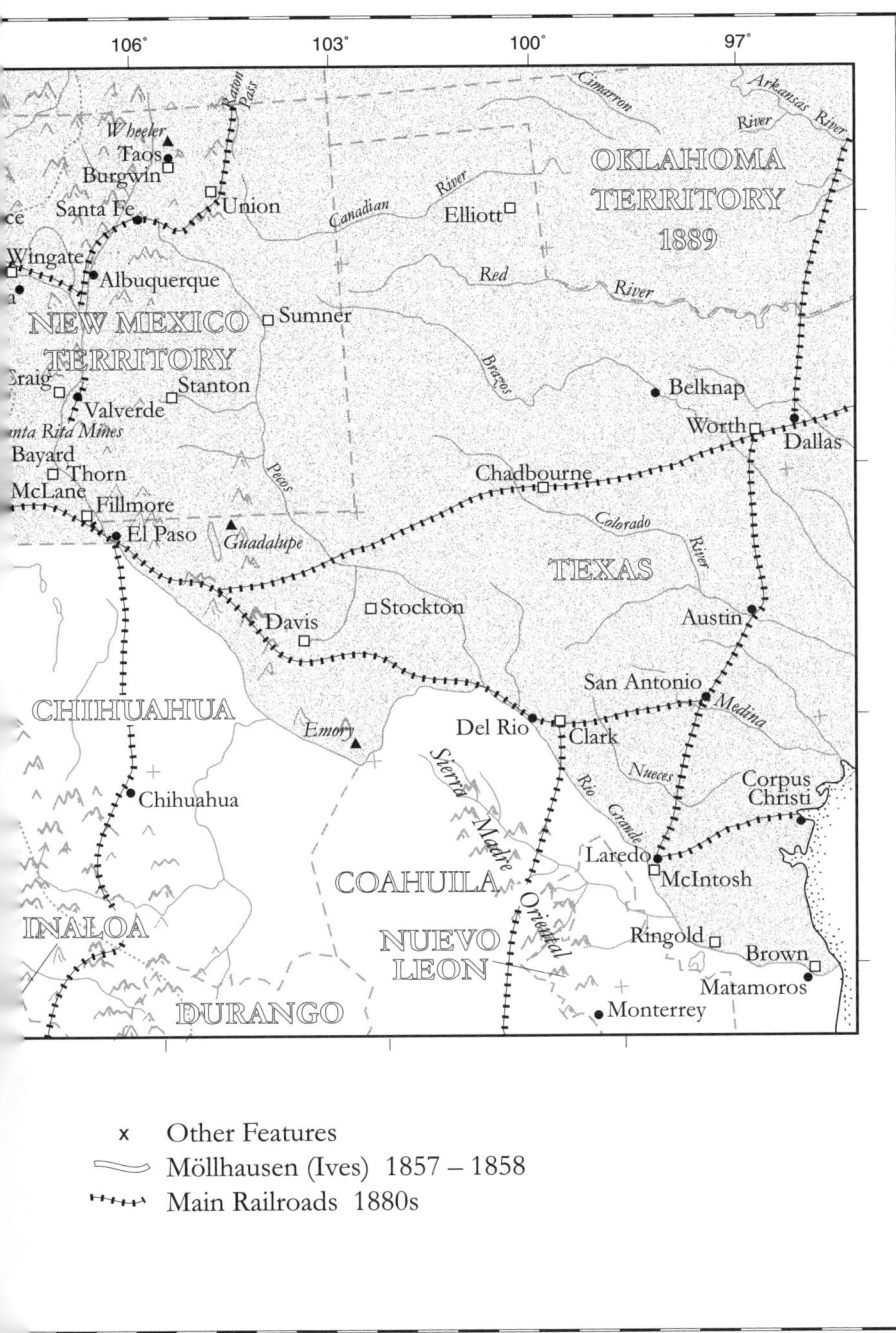

106°   103°   100°   97°

Cimarron

Arkansas River

River

Wheeler
Taos
Burgwin

Santa Fe □ Union

Canadian River

Elliott □

OKLAHOMA
TERRITORY
1889

Wingate

● Albuquerque

Red River

NEW MEXICO
TERRITORY

□ Sumner

Brazos

raig □

Stanton

Belknap ●

● Valverde

Pecos

Worth □

Santa Rita Mines

Bayard

□ Thorn

Chadbourne □

Dallas

McLane

Fillmore

● El Paso ▲ *Guadalupe*

Colorado River

TEXAS

Davis □

□ Stockton

Austin □

San Antonio

Medina

CHIHUAHUA

Emory ▲

Del Rio ● □ Clark

Nueces

Corpus
Christi

● Chihuahua

Sierra Madre Oriental

Rio Grande

Laredo ●

□ McIntosh

INALOA

COAHUILA

Ringold □

Brown □

NUEVO
LEON

● Monterrey

Matamoros

DURANGO

x   Other Features

〰   Möllhausen (Ives) 1857 – 1858

⊢⊢⊢⊢   Main Railroads 1880s

The Southwest Borderlands, 1854–1900 (Map by Susan Alta Martin) to prepare for a gust of wind, and a heavy rain, which he thought betokened by a rushing noise he heard. . . . We landsmen from the interior, and unaccustomed to such movements of the water, stood contemplating with astonishment the rush of the tide coming in from the sea. . . . In twenty minutes the place where we lay asleep . . . was three feet under water and our blankets were all afloat."

While waiting for the *Explorer,* Möllhausen arrived at Fort Yuma via Fort Tejon in sufficient time to do some collecting. Of the fifty-five species taken on the entire trip, only a listing is given, but they arrived in Baird's hands in time to be incorporated in volume 9 of the *Pacific Railroad Reports* (Baird, Cassin, Lawrence 1858). With winter at hand, many migrants, as well as residents, were secured in the vicinity of Fort Yuma. Birds new to Möllhausen were the Crissal Thrasher, "the second specimen ever collected," and American Pipit, undoubtedly collected on fluctuating mud flats of the Colorado River (Ives 1861). Over thirty years later, C. Hart Merriam (1890) was the first to note these pipits probably breed near the "timber line on San Francisco Mountain" south of the Grand Canyon.

On January 9 the *Explorer* arrived and joined the three detachments already at Fort Yuma, which, as Ives (1861) wrote, was the "Botany Bay of military stations." In two days the steamboat was again underway, with an escort for protection, while a supporting pack train followed overland. The problems of navigating the river, especially that year, were indeed difficult, for the Yuma Indians had "never seen it so low." Almost immediately the steamboat ran aground within view of the jeering onlookers of the fort, a situation that was repeated countless times up the river.

Leaving Fort Yuma (elevation, two hundred feet), the slow, steady course upstream on the "opaque" silt-laden river brought seemingly endless hidden surprises of shoals, sandbars, and rock formations with spectacular "varieties of coloring."

Indians were encountered at many points along the river, and in Chemehuevi Valley, Ives (1861) wrote with some humor: "Mr. Möllhausen has enlisted the services of the children to procure zoological specimens, and has obtained, at the cost of a few strings of beads, several varieties of pouched mice and lizards. They think he eats them, and are delighted that his eccentric appetite can be gratified with so much ease and profit to themselves."

Along the course of the river, Möllhausen secured a number of species, several of which were not collected on the Whipple survey four years earlier. Among them was a large, long-legged Wood Stork that was referred to as "Colorado turkey . . . because [it] is said to be abundant on the Colorado river, especially about Fort Yuma" (Baird, Cassin, and Lawrence 1858).

Passing the Needles, they entered the "profound chasm" of Mojave Canyon where "a solemn stillness reigned in the darkening avenue broken only by the splash of the paddles or the cry of a solitary heron startled by our approach" (Ives 1861). On reaching 380 miles (elevation, approximately nine hundred feet) above Fort Yuma, they terminated their steamboat segment of the expedition on March 8.

While waiting the arrival of the supply train, Ives explored upstream in a skiff for several days and then sent the *Explorer* downriver. On March 23, Ives and the remainder of the party, which totaled twenty-five men, set off east to the Hualapai and "Moqui" (Hopi) villages, passing Bill Williams Mountain and the San Francisco Peaks before reaching Fort Defiance on May 23, 1858.

While on this unusually diversified trip, Möllhausen illustrated for Ives' report the magnificent grandeur of the lower Grand Canyon, along with the local Indians, and some of the activities of the party. Newberry, the first geologist to visit the region, expounded about theories that still stand on the origins of the stupendous canyon and its many tributaries. The expedition ended none too soon, for later the same year, hostilities broke out between the army and the Navajo Indians, which lasted several years. Shortly after completing this expedition, probably one of the most exciting and atypical adventures of the Southwest, Lieutenant Ives resigned his commission and joined the Confederacy, where he served on the staff of Jefferson Davis.

AUDUBON OF THE PACIFIC, 1847–1869

֍ IN 1846, ANDREW Jackson Grayson (1818–1869), like many adventuresome young men, undertook to fulfill an appealing idea of the day—to organize a caravan and travel from Louisiana to California. Upon reaching Fort Bridger, some members split from his group to join the ill-fated Donner Party. After five and a half difficult months, Grayson, with his wife and infant son, finally arrived with his party in Yerba Buena (San Francisco). Shortly thereafter, he entered into a series of business ventures and also received a commission of colonel to serve in John C. Fremont's battalion.

In 1853, at the age of thirty-five, Grayson took another profound step, this time in the direction of art and science. Although he had an early interest in nature, his transition was not brought about until he and his wife leafed through

Audubon's *Birds of America,* recently acquired by the Mercantile Library Association in San Francisco. Realizing the famous artist and naturalist had not illustrated the birds of the Pacific region, Grayson determined that by following in a similar manner, he might undertake this enormous task. During the remaining short but remarkably productive sixteen years of his life, he became a self-taught artist and naturalist, showing unusual ability in both areas. His vivid watercolor portraitures of birds in natural surroundings include intricate details of plants and insects similar to Audubon's. However, Grayson painted his subjects with more realistic postures, a decided advancement over some of Audubon's work. Grayson also wrote accompanying species accounts for the birds he portrayed, earning him the title, "Audubon of the Pacific" (Bryant 1891a). Over half of the species he described occur along the Southwest border region.

In 1856, Grayson's contact with the Smithsonian Institution brought him into correspondence with Spencer Baird, who encouraged him to collect birds and contribute observations on their habits. His specimens from the coastal central California region are repeatedly mentioned in the *Pacific Railroad Reports* (U.S. War Department 1855–1860).

In 1858, Grayson, a restless man, always seeking to improve his monetary position in an effort to paint birds, traveled to a mining prospect along the Colorado River near Fort Yuma. The venture was unsuccessful, but he did paint a family grouping of Gambel's Quail. With attempts to gain complete financial independence continuing to elude him, Grayson traveled twice to Mexico before he and his wife moved to Mazatlàn in 1859.

While in Mexico, Grayson's correspondence with the Smithsonian Institution increased. Baird encouraged him to collect more specimens, while Grayson, in return for his services, requested funding. Disappointed with no response, Grayson, growing more anxious for financial support, then attempted to interest the Smithsonian in his bird paintings. Neither request was ever answered, despite the fact that he continued to collect for Baird.

During this period, Grayson's folio of Mexican birds increased. In 1865, discouraged in his attempts to interest Baird, he turned to the Mexican ruler Napoleon III, Archduke Maximilian of Austria, and the Mexican Academy of Science and Literature. He received some encouragement and the promise of financial assistance from Maximilian, but that proved short-lived, for Maximilian fell from power and was executed in 1867.

Gambel's Quail, Callipepla gambelii *(Gambel), 1843. Painted by Andrew Jackson Grayson, 1847–1869. (Stone 1986; courtesy of the Bancroft Library University of California, Berkeley)*

PLATE 35
GAMBEL'S QUAIL
Callipepla gambelii (GAMBEL)

Grayson was very active while in western Mexico, also collecting on the islands of Socorro, Tres Marías, and Isabella. In 1867–1868, he painted and described two birds endemic to the island of Socorro, naming them Socorro Towhee and Socorro Wren (Taylor 1951).

In addition to painting, Grayson carefully wrote detailed life history sketches of birds, from which this segment on the Inca Dove is taken: "It is regarded by the Mexicans with feelings of tenderness, and their mildest expressions of love are referable to this little dove—as *mi palomita,* my little dove. They exhibit the most ardent attachment for their mates and may be often seen caressing each other in a loving manner. But with all their innocent looks, they are possessed of violent paroxysms of jealousy, and a pugnacity equaling the game-cock" (Bryant 1891a).

When French naturalist René Primevère Lesson named the Inca Dove in 1847, he may have been uncertain about the origins of the Indians in the Western Hemisphere, for he named this bird in honor of the Andean Incas of Peru, rather than for the Aztecs of Mexico, where it commonly occurs. Acting assistant surgeon H. B. Butcher (1868), while serving at Fort McIntosh at Laredo from 1866 to 1867, noted the first occurrence of the Inca Dove north of the Rio Grande. The slow northward extension of this small, scaly-looking dove was coincident with the human development of the region beginning in the 1870s (Monson 1954).

Grayson was disadvantaged from the start in gaining recognition as a bird portraitist, but his passion was undaunted. His isolation from the ranking ornithologists of the period, and his lack of associates, mentors, and formal training in science were serious obstacles. These problems seemed of minor consequence, however, compared to being robbed twice by Mexican bandits, surviving two shipwrecks in Mexican waters, a continual reoccurrence of yellow fever, and his son Edward being killed in the streets of San Blas. Only after he contracted another serious fever on Isabella was his work finally halted, when he suddenly died. The Socorro Dove *(Zenaida graysoni)* was named for his son Edward by G. N. Lawrence (Hood 1933).

The impact of Grayson's work in western Mexico, particularly on insular or island forms, can be noted by the great number of specimens he sent to the Smithsonian Institution, and one that Lawrence named in his honor: the Socorro Mockingbird *(Mimodes graysoni)*.

Through all of these adversities, Grayson's devoted wife Frances remained by his side and, following his death, she sought publication of his paintings. Failing this, she donated the paintings to the University of California at Berkeley in 1879. In tribute to her, Baird (1865) named in her honor the Tres Marias Chat *(Granatellus francescae)*, adding "I cannot more appropriately dedicate it than to Mrs. Grayson, to whose encouragement Col. Grayson owes so much of the persistency and success with which he has prosecuted the study of the ornithology of California and western Mexico."

Baird also wrote Mrs. Grayson concerning her husband's work: "I consider the memoir, next to the work of Mr. Audubon, the most important contribution yet made to American ornithology, in the form of good illustrations and interesting biographies. The plates are drawn with exceeding care, and represent the minutest features of the species; while, as regards the fidelity of coloring, I do not believe they have ever been excelled" (Palmer 1928b).

Sadly, the only full-sized lithograph published during Grayson's lifetime was that of the California Quail. But several of his remarkable prints appear in the "Distributional Check-list of the Birds of Mexico," parts 1 (Friedmann, Griscom, and Moore 1950) and 2 (Miller et al. 1957). Fortunately, Grayson's paintings and writings were ultimately entrusted to the Bancroft Library, from which Lois Chambers Stone assembled his work and biography into *Birds of the Pacific Slope* (1986). Appearing in two parts, his spectacular folio contains 157 individual prints (19″ x 25″) and is separate from the companion descriptive work. Together, this remarkable set is a lasting tribute to this determined artist-naturalist. It is especially important because it ranks him in history, and rightly so, among the most talented ornithologists of North America.

THE TROCHILID PROFITEER, 1851

By the mid-1800s the fascination for private bird collecting was rather common, and hummingbirds, the tiny jewels of the New World, became especially popular. Many men saw opportunities to profit by collecting skins for ornithologists and museums in addition to the commercial market.

French bird fancier Adolphe Boucard (1839–1904) fit this mold of individuals who not only collected for prominent scientists, including Philip L. Sclater, but conducted a business of supplying skins and feathers for the millinery trade. In 1851, while living in San Francisco, he explored several schemes for a commercial trade in hummingbirds. "I also collected many species of birds, and more particularly Humming-birds [Anna's and Rufous]. . . . I found many nests . . . and I had as many as sixty of them alive, all taken from the nests. I fed them with fresh flowers and small insects. Some of them lived four months. . . . I succeeded in keeping them alive and well for a long time. My intention was to send them alive to Europe, but even the most robust died at sea, and it was a complete failure" (Kofoid 1923).

In support of his commercial motives and in an effort to encourage the trend, Boucard wrote: "Even supposing that the fashion would continue for ever, it is my opinion that certain species of Birds are so common that it would take hundreds of years before exhausting them." Little wonder the Royal Society for the Protection of Birds was founded in England by irate ladies objecting to the millinery trade in 1890.

THE GUANO HUNTER, 1856

𝒯HE FIRST IMPRESSION of many observers of islands in the Gulf of California in fall or winter might be loneliness and desolation. However, on Isla Raza in early spring, the scene undertakes a sudden transformation, as several thousand gulls and terns return to breed as they have for centuries. Two species of terns, Royal and Elegant, arrive and congregate tightly in an exposed low basin and form a solid white mass, which glows against the darker sand and rocks. Heermann's Gulls, their predominate predator, also nest and surround them in an even distribution throughout much of the remainder of the island.

In 1855, Federico Craveri (1815–1890), a widely traveled Italian educator and chemist living in Mexico, was asked by the Mexican government to investigate the potential for mining guano on the Pacific and Gulf islands surrounding the Baja Peninsula. The next year Craveri began a lengthy trip through the region and collected many birds, including a new alcid from Isla Raza (Cooke 1916). This specimen was eventually lodged undescribed in the Zoological Museum of Turin University in Italy. In 1865, Count Tommaso Salvadori realized, while reviewing Craveri's collection, that it was a new species and named it Craveri's Murrelet *(Synthliboramphus craveri)* (Palmer 1928b).

While on an extensive voyage in the Pacific Ocean in 1874 and 1875, Dr. Thomas Hale Streets (1877) discovered on Isla Raza the first nest of this murrelet "breeding in holes in the rocks, amid the innumerable gathering" of Heermann's Gulls. The much smaller murrelet, half the size of the gull, must depend on crevices for protection from the larger predator.

Approximately fifty miles south of Isla Raza is a massive Midriff monolith, Isla San Pedro Mártir, which rises with tremendous bluffs above the sea. Whitish seabird excrement covers the entire island, including the predominate cactus, the giant Cardón. Over three decades after Craveri's visit to the region, Col. Nathaniel S. Goss (1826–1891), one of the original members of the AOU, reported that a company with about 135 "Yaquie Indians" was removing guano from this island. Goss (1888) noted that although the men were "continually disturbing and often robbing the birds . . . the Indians are not as destructive as the white race, and as the Company feeds them all, [they] seem to care but little for the eggs." Continuing, he added that "when undisturbed thousands upon thousands [boobies] will breed there." Historically, egg collecting on the Midriff islands by mainland natives has been a serious threat (Anderson, Mandoza, and Keith 1976; Bahre 1983).

THE XÁNTUS TRADITION, 1857–1863

🌀 ABOUT 1850, FOLLOWING the Hungarian war for independence, John Xántus (1825–1894), expelled from his homeland, emigrated to America. After casting about and enduring many trying circumstances and, not infrequently, fantasized adventures, he gained his U.S. citizenship by joining the U.S. Army in 1855. The "Xántus tradition" refers to a rather cursory review by Harry Harris (1934) of the "almost mythical nature" of the "American adventuring" of this early naturalist, which was based primarily on Xántus's memoirs in two Hungarian books. Harris's review also includes a brief summary of the massive collection entries sent by Xántus to the Smithsonian Institution, notes on unpublished letters, and an undated complimentary memorandum of his contributions to the Smithsonian written by Spencer Baird to the president of the Hungarian Academy of Science.

As for the outstanding fieldwork of Xántus, the naturalist, there can be no doubt. Much of his early history, as presented in his two books, however, was at least suspected to be untrue by Harris. But it wasn't until Henry Miller Madden (1949), Xántus's biographer, failed to corroborate substantial segments of the books and realized they were often fabricated or plagiarized material, that they were repudiated as "pure fiction." In a similar vein, biohistorians Edgar E. Hume (1978), Ann Zwinger (1986a and b), and Barbara and Richard Mearns (1992) reveal Xántus to be a hypersensitive, jealous, and boastful individual, quick to learn, but with an extremely abrasive personality.

When considering that much of Xántus's history is in question, it should be noted his entry into the Southwest is thoroughly documented by his correspondence with Baird. The Xántus letters, safely preserved in the Smithsonian Archives, were retrieved by Zwinger (1986a and b) and provide unabridged authenticity. They are a wonderful resource, not only in revealing much of his personality and behavior, but also in providing an insight into the life of an enlisted man and, most notably, into the natural history of the area where he collected.

While on his first duty at Fort Riley, Kansas Territory, Xántus learned how to improve his specimen preparation from Dr. William A. Hammond, who later became the first surgeon general to achieve the rank of brigadier-general. Hammond, recognizing his talents, arranged with Spencer Baird for the enthusiastic young soldier to be transferred to Fort Tejon in southern California.

In 1854, one year following Lt. Robert Williamson's railroad survey

John Xántus (1825–1894).
(Hume 1942)

through the region, Fort Tejon was established to secure the mountain roads of the vicinity. Situated in a narrow pass (4,239 feet), the fort lies between the southern end of the great San Joaquin Valley and the eastern extremity of the Mojave Desert, an area of significant seismic activity. Giant oaks occur in the pass and, in the nearby mountains, which rise over eight thousand feet, conifer forests abound.

In June 1857, Xántus arrived at Fort Tejon with a rank of sergeant to perform the duties of hospital steward. In his first letter to Baird, he wrote of his "great mortification" that some of the staff upon which he was to depend had rotated out and the assistant surgeon in charge of the hospital "dont takes any interest at all in NatHistory." About the conspicuous large mammals, he noted: "We have here grizlys [Grizzly Bear] in great abundance, you cannot walk out half a mile, without meeting some of them, and as they just now have their clubs [*sic*], they are extremely ferocious to, I was already twice driven on a tree, and close by to the fort."

By the end of June, Xántus was disillusioned with all the personnel at the fort, stating "everybody is a gambler and drunkard, they sit day & night in whisky shops, or gambling holes; and instead of supporting me they ridicule me and throw every obstacle in my way." In his prideful manner he added, "I treat them of course with princely contempt, and go on with doubled step in my path." It must have been extremely difficult for him, as an enlisted man with no real authority, an unusual avocation, a heavy accent, and an extremely proud and arrogant attitude, to achieve his many accomplishments.

The next month Xántus wrote that he had a "menagerie," which included a "fine grizzly cub" that somehow had "tore to pieces already the Colonels dog," no doubt adding to his troubles.

His collections continued with many "alcoholic specimens" lost because "the several severe Earthquaque shocks broke almost everything here." In August, in need of liquid preservatives, he gave instructions to Baird that "should [you] send alcohol, put the key in a box, & fill the space with straw, this manouvre will avert thirsty peoples attention, and always safely arrive at the destination."

In November, Xántus noted rumors circulating about the fort that gave "strong reason to anticipate the entire abandonment of this Post in consequence of the Earthquaque shocks, & therefore I asked you . . . to procure a suitable transfer." His insistence in asking Baird for a transfer continued unabated through his remaining stay and was, perhaps, based primarily on his inability to get along with his superiors.

Among the many requests that Xántus made of Baird were books and reports, which Baird often sent, but like many other items, these were long in coming or never received. Almost three months after his arrival at the post, Xántus wrote "I have to inform you Sir on a new misfortune, which will show you, that I am really persecuted by fate in every respect.—The Ordinance Sergeant . . . rode down to Los Angeles . . . to bring up my mail. He found . . . package with Washington stamp . . . put althogether in a handkerchief, and placed on the pommel of his Sadle. Taking a farewell dram—as the P. O. is barroom also—the horse got mad, broke the halfter, & run away. When cought (about 10 miles out of the town) was everything on her back, except—my unfortunate mail."

A few months later, Xántus received Sitgreaves' account of exploring the Zuni and Colorado Rivers, whereupon his ridicule extended to the naturalist of that expedition. "By the by I see in Woodhouse [(1853)] that he found this bird [Townsend's Solitaire] quite abundant [Woodhouse stated "exceedingly"] in N. Mexico. It may be, but Mr Woodhouse found every bird quite abundant, *extremely common, very common,* etc there; which indeed must be a rich field. I presume, he brought home several thousand accordingly."

All the while, Xántus was vigorously collecting and sending specimens not only to Baird, but to Philadelphia. "I am much vexed, that . . . I contributed so extensively to the Museum of the Pha. Academy, and they never even noticed it. I sent only to Cassin over 150 species of birds, & you cannot find a line of acknowledgment in the Proceedings."

Xántus was receiving some recognition, however, for he had a visit from Baldwin Möllhausen, who had accompanied the Whipple railroad survey. Returning from Europe, the German naturalist was again on the west coast preparing to embark with Lt. Joseph C. Ives on the steamboat exploration of the Colorado River. Xántus wrote Baird about the experience:

> He informed me that he heard much in Los Angeles about my collection, and come up here expressly to see it.—I confess you sincerely that I felt since some time uneasy on hearing, that Möllhausen is coming & is going to collect (so to say) within gunshot of my kingdom. . . . I do not fear him, but on the contrary intend to assist him in procuring specimens at my own ground.—He was completely beaten by seeing my collection, & he confessed to others this morning, that he passed a sleepless night, and never will attempt to collect here anything, although he came with the intention, to do it.

He went on to add "*confidentially*" to Baird that Möllhausen "tried to induce me, to sell my collection to the King of Prussia, garanteing that he will confer a great favor on me, especially A. Humboldt. I of course emphatically refused . . . in the presence of several Officers, amongst them Dr TenBroeck," who was Xántus' immediate superior.

From the moment Xántus arrived at Fort Tejon, he was in want of his favorite "shot gun" which, along with many personal effects, was never received from Fort Riley. He substituted various weapons until another "great calamity. Lt. Beall [Edward F. Beale (1822–1893)] with his camels passed here several days ago, & having no escort he enlisted citizens, & armed them from the ordnance . . .

he took every musket, Carabine, pistol, & sabre on hand; and I am now entirely naked (scientifically speaking)!"

Following that episode, just prior to Xántus receiving a new gun from Baird, the commanding officer prohibited the firing of guns within the garrison. Later ignoring the order, Xántus, after being caught and reprimanded, wrote sarcastically that "the Major rules over us with his arbitrary—Turkish Pasha like—power, & arbitrary will."

Xántus apparently did not alienate himself to everyone, for he took a liking to another emigrant, a German sergeant named John Feilner (?–1864), who was also stationed at the fort. He wrote Baird that the sergeant was a "pupil of mine" and that every effort should be made to accommodate him as a serious collector. In 1860, Feilner, a year before his commission and four years before he was killed, added the first Flammulated Owl to our fauna in northeastern California (Baird et al. 1874).

Xántus was recently credited for discovering Cassin's Vireo (*Vireo cassinii*) which, along with the Plumbeous Vireo, was split from the Solitary Vireo in 1997. In conjunction with this vireo, he was also supplying nests and eggs of other birds to Thomas Brewer. Securing them was sometimes difficult. In this case, he took "fully four days meditating & devising plans, as how to obtain it." After the troubles and dangers of finally reaching it, he wrote lightheartedly that he "had many adventures of very similar character, the description of all together would make a very amusing volume, & might be entitled '*The Rambling & climbings of a fool*.'" Xántus was very insistent in expressing his desire to write "some memoir" on the birds he collected. In depending on Baird to determine the species validity of those sent, he also expected in return, a reasonable qualification. "If you find some new species amongst my birds . . . and you liked to introduce them in your progressing work on birds; I have no objection provided you introduce my description. But in order to do this,—I suppose—you ought to publish in my name before, in the proceedings of the Academy, or somewhere else. In this case, I wanted to name the first bird '——— *Hammondii*.' In honor of my excellent friend Dr W. A. Hammond U. S. A. And the Second '*Bairdii*' in your honor sir—provided you have no objection." Baird abided by his wishes and placed in the Proceedings under the name of John Xántus de Vesy (1858) a description of Hammond's Flycatcher (*Empidonax hammondii*). His second request was humbly acknowledged by Baird but was never fulfilled. He occasionally sent sketches with his specimens.

Among the specimens Xántus sent to Baird were two he believed to be

Yellow-bellied Flycatchers. Baird, however, with past experience of describing this species, was still having some difficulty. Recognizing "distinct" features such as plumage colors and quill lengths, he wrote: "In view of all these circumstances, therefore, it may be well to give it provisionally a new name, and none would be more appropriate than that of *E. difficilis*" (Baird, Cassin, and Lawrence 1858). Previously named the Western Flycatcher, it was split into two species in 1989. The birds of the extreme west retained their specific name but have been given the common name of Pacific-slope Flycatcher, while those birds of the Rocky Mountains became the Cordilleran Flycatcher.

Xántus's (1860a) second paper was a "Catalogue of Birds Collected in the Vicinity of Fort Tejon, California, with a Description of a New Species of Syrnium." It comprised a listing of 144 species collected over "about 17 months" in which he briefly compared species numbers with Dr. Thomas C. Henry's work at Fort Thorn along the Rio Grande. The species of Syrnius he discovered was the rare Spotted Owl, which he secured in 1858.

In anticipation of adding the California Condor, or its eggs, to his collection, Xántus wrote "I cannot find in Audubon an immense Vulture, which leaves here. . . . They are quite numerous, but as yet I have no specimens, being entirely confined to the high mountains; but I am informed some of them measure from tip of wing to tip of wing fully 18 feet [actual wing span is 8 1/2 to 9 1/2 feet]. I think this statement cannot be much exaggerated, as I myself often mistook them on hills for mounted men. I liked to know the name of these birds, I will secure some soon no doubt." More on the life history of these great vultures was detailed a half century later (Finley 1906, 1908; Sharp 1907).

Of incidental interest, exactly one hundred years after Xántus observed his first California Condor, the author, out of breath after topping a limestone ridge only four miles southwest of Fort Tejon, looked up in ecstatic awe to view for the first time this majestic vulture, resplendent in adult plumage, wheeling directly overhead less than seventy-five yards away. Since that memorable day in 1957, I observed intermittently the declining remnant wild population for twenty-eight years until, in their last year of freedom, I was fortunate to view three birds twenty-five miles west of Fort Tejon in 1985.

The discussion between Baird and Xántus was ongoing regarding possible transfer locations. While Xántus was continually reminding Baird of his plight, Baird, no doubt, understanding his complex personality, responded with several potential solutions. One was a move to Fort Yuma to which Xántus replied "I must positively I decline my transfer to that sandhill, by many reasons. The medical officer of that Post is a half crazy quarrellsome litle boy, a certain Dr [George] Hammond; and the Comdg Officer a very bitter enemy of all my operations as well, as my humble person."

Baird continued to look for other remote and unexplored places to satisfy Xántus, including the possibility of the Cape region of Baja California. Xántus was subsequently discharged from the army and employed as a tidal observer for the U.S. Coast Survey on the southern tip of Baja California. After a brief training stint in San Francisco, he was already experiencing difficulties with his immediate superior. Nevertheless, in 1859 he sailed down the coast and arrived at Cape San Lucas, his home for the next twenty-eight months.

The first three months were productive enough for Baird (1859b) to publish a paper on the forty-two bird species that he had received from the naturalist. Included on this list were two new species that Xántus (1860b) described. The first was the Gray Thrasher, which he noted as "very abundant at the Cape, and its nests are found among the cactuses in large numbers." The second was a "considerably weatherbeaten" specimen taken fourteen miles off the Cape. Although Xántus applied its specific name, it was later given a common name of Xantus' Murrelet.

Before the year was out, Xántus sent a new hummingbird to Baird, who passed it to George N. Lawrence. After describing it, Lawrence then dedicated it in honor of its discoverer in 1860. But when Xántus learned about the designation, he indignantly reminded Baird of their previous arrangement. "I am much obliged to Mr Lawrence that he named the humming bird Xantusii, although I had preferred to describe myself. You know very well we agreed, that the birds shall be described all by me, or in my name, should you prefer not to await my arrival there" (Zwinger 1986b).

Baird later wrote regarding Xantus' Hummingbird (*Hylocharis xantusii*): "This is a new and well-marked species, and although belonging to the North American fauna cannot be claimed for the United States, having thus far been only taken at Cape St. Lucas by Mr. Xantus" (Baird, Brewer, and Ridgway 1874).

Xántus was still languishing over Woodhouse (1853) and, in this case, his visual account of the "Rock" or White-throated Swift at Inscription Rock. This event reminded him of when he was only able to glimpse an unknown picid. Giving Baird a somewhat confusing description, he commented with mordant wit: "I am sorry I cannot say more positive about this bird. . . . I wish I had this time the eyes of one of our American naturalists, who accomplished notoriously such feat, as to name even & describe a swallow on wing, and that most minutely, not forgetting even dimensions of tail, tarsus, feet etc. of course the swallow was never before or after seen" (Zwinger 1986b).

In June 1861, Xántus spent several weeks with Andrew Grayson and his wife in Mazatlàn. After meeting Grayson two years earlier in San Francisco, Xántus wrote Baird that he "showed me a letter to him from you, where you stile him 'the Audubon of California'—The Colonel . . . does not know to distinguish a flycatcher from a dove." Grayson, however, was very complimentary about Xántus to Baird until several years later. Then, in frustration that no financial assistance was forthcoming, he remarked "Xántus had ample facilities furnished him. . . . He never went into the woods himself, but hired natives to shoot his birds" (Stone 1986).

In February 1861, Xántus, having endured privations in the extreme, "chubascos," hurricanes, gales, drought, intense heat, and isolation with limited companionship and assistance, complained to Baird that no source of news had been received for at least four months. "I am quite sick indeed of this place, every day seems a long year, and every one with the same monotonous desolation around me, not affording the least pleasure, variety, or enjoyment of any kind. . . . I wonder what all happened since in the world!!" To this he added a postscript: "I am now of Gods grace nearly two years perched on this sand-beach, a laughing stock probably of Pelicans & Turkey buzzards, the only signs of life around me" (Zwinger 1986b).

Over eighty years later, John Steinbeck (1951), on his voyage aboard the *Western Flyer* to the Gulf of California, even remembered Xántus's "broadening" reputation. While ashore at the Cape, he remarked to the local hotel manager that Xántus might have "brooded and wished" for accommodations such as these while on his ambitious adventures. To this the manager replied: "'Oh, he was even better than that.' Pointing to three little Indian children he said, 'Those are Xantus's great-grandchildren,' and he continued, 'In the town there is a large family of Xantuses, and a few miles back in the hills you'll find a whole tribe of them.'"

In the fall of 1861, Xántus was recalled and discharged from the Coast Survey. After Xántus made a trip to Hungary, Baird managed to obtain an appointment for him as consul of Manzanillo. In 1863, after assuming the post, his difficulties commenced almost immediately, and he was subsequently dismissed. He prolonged his stay long enough to obtain a sizable collection of mainland birds.

Xántus was controversial in almost every activity that he was engaged in throughout much of his life. Nevertheless, despite his many faults and failings, this energetic naturalist, intrigued by natural history, was a major participant in advancing knowledge about our native fauna.

5 FRONTIER UNREST

"my enthusiasm runs so high, that sometimes as I stand alone in the
wilderness, thousands of miles from home and friends, hot, tired, breathless with pursuit, but holding in my hand and gloating over some new or
rare bird, I feel a sort of charitable pity for the rest of the poor world, who
are not ornithologists, and have not the chance of pursuing the science
in Arizona."
—ELLIOTT COUES (1865)

DRAMATIC HUMAN CHANGES continued in the region. In 1862, the
bloodiest battle of the Civil War in the Southwest was fought at Valverde along
the Rio Grande. In the same year, a Confederate force penetrated northward to
Santa Fe, where it was defeated at Glorieta Pass before withdrawing to Texas.
The next year Arizona was separated from New Mexico near the 109° longitude and became a Territory. The army was engaging the Indians more frequently by this time: the Mescalero Apache had just accepted their fate and
were forced from their traditional mountain homelands in south central New
Mexico and west Texas to Bosque Redondo along the Pecos River; Col. "Kit"
Carson (1809–1868) was "ordered" to sweep the Navajo from their spectacular
canyon lands to the same dismal site with the Mescaleros; Mangas Coloradas
(ca. 1795–1863), chief of the Eastern Chiricahua, was brutally murdered as a
prisoner at Fort McLane; and the several tribes and bands of Apache remained
openly at war in western New Mexico, eastern Arizona, and adjacent Mexico.
However, they too were relentlessly pursued by Gen. George Crook (1828–1890)

until they were primarily compressed into areas centering on Fort Apache in the White Mountains and San Carlos along the Gila River. In the extreme southeastern corner of Arizona Territory, Gen. Oliver O. Howard (1830–1909) succeeded in establishing the short-lived Chiricahua Reserve in 1872.

RUNNING THE BLOCKADE, 1863–1864

IN 1863, HENRY Eeles Dresser (1838–1915), an English businessman and trader, arrived in Texas with a shipment of goods for the Confederate government. His wide range of experiences included many activities that centered on his favorite study—ornithology. He was a prominent author and a member of the Linnaean Society, the Zoological Society of London, and for a period of six years secretary of the British Ornithologists' Union.

Dresser (1865, 1866) was among the first naturalists to compile an annotated list of the birds he observed and collected near San Antonio, the coastal regions, and the lower Rio Grande Valley, including the adjacent area of Matamoras, Mexico. Of the 272 species he noted, a few were sent to him from Fort Stockton. On his return to England, he published his notes in the English ornithological journal *Ibis*. Up to this time, Elliott Coues (1878) considered him to be "one of the chief authorities for this locality."

Dresser (1865), while exploring the Medina River near San Antonio, was the second naturalist to secure the rare Golden-cheeked Warbler and the first to add it to our fauna north of the border. In 1878, George H. Ragsdale secured a second specimen north of the border in Bosque County (Purdie 1879), and W. H. Werner found four nests, all situated in junipers in Comal County (Brewster 1879b). Dresser also noted the White-eyed Vireo as common in west Texas.

At Matamoras, Dresser (1866) observed another rare bird, the stately Whooping Crane, and the beautifully plumed Roseate Spoonbills wading about the lagoon. Even at this early date, he recorded the decline of this striking pinkish marsh wader, adding "in June 1864 I saw two or three on Galveston Island, where it is known under the name of 'Flamingo.' I was told that it had bred on the island in former years, but it does not do so now, being too much disturbed."

The Mississippi Kite, an elegant raptor of grace and buoyancy, was "by no means an uncommon bird in Texas" during Dresser's visit. Although Dresser found them nesting near San Antonio, Samuel Woodhouse secured three specimens in 1851 in what is now New Mexico. This kite was not noticed by natural-ists further west until almost a century and a quarter later when it was found nesting in southeastern Arizona (Levy 1971; Carothers and Johnson 1976).

Witmer Stone (1916a), in contact with Dresser regarding the tragic death of Adolphus Heermann, reminisced kindly about the visiting English naturalist, stressing "his cheerfulness and sweetness of temper, qualities which even those who knew him as did the writer, only as a correspondent, can readily appreci-ate."

A HOT-TEMPERED IRISHMAN, 1857–1861

IN 1857, FORT Buchanan was established near the headwaters of Sonoita Creek, just over forty-three miles southeast of Tucson, New Mexico Territory, and twenty-four miles north of the Mexican border, in an effort to protect the travel routes through the region. Considered to be one of the most unhealthy posts in the Southwest, it was evacuated and burned four years later, before the arrival of a Confederate unit in 1861.

Assistant surgeon Bernard John Dowling Irwin (1830–1917), aggressive, observing, and sometimes outspoken, served at Forts Union and Defiance before arriving at Fort Buchanan shortly after its establishment. Capt. Richard S. Ewell (1817–1872), an acquaintance and post commander, referred to Irwin as "a red-headed, hot-tempered Irishman" (Serven 1965).

As post surgeon, Irwin, very critical of Fort Buchanan so near a "swamp morass," wrote: "If, in the selection of the site for the post, any attention had been directed to the local character of the place, it would have been apparent at a glance that it would prove unhealthy when occupied as a garrison. It is needless to add that no medical officer was consulted on the subject" (Serven 1965). The structures were only a series of "temporary jacal buildings" consist-ing of "pickets" in which the siding and roof "chinks" were "filled up with mud," affording in wet weather "dirty shower baths to the unhappy occupants."

The fort was surrounded on three sides by a "*cienega,*" which Irwin strongly felt was the problem. He added "the troops have suffered continually from malarial disease, which has attacked every person at the post during the last year." But, in spite of these terrible conditions, he considered the general "climate of Arizona very healthy, and particularly pleasant."

Irwin was linked to several noteworthy events during his tenure at the fort. In 1858, the Overland Mail Company began service and, in the Dragoon Mountains, three Mexican employees axed four of their associates. Upon learning of the incident, Irwin set out immediately on a 115-mile ride to provide medical assistance. Only Silas B. St. John (1835–1919) was saved, by Irwin's immediate amputation of his arm (Thompson 1935).

The birds of the area were also of interest to Irwin. "Much might be written about the rare and beautiful birds that abound in this country, many of which are remarkable for the gorgeous beauty of their plumage" (Serven 1965). He was undoubtedly aware of the Mexican Boundary Survey and its leader when he listed "Emory's hawk" among the many birds he "met with." To what species he was referring has not been mentioned in the literature, nor was it recognized by scientists of the period.

It would certainly be speculative to note which publications on southwestern ornithology, if any, were accessible to him at this isolated outpost. The result was that when he listed birds, he applied descriptive terms such as "hooting owl" for Great Horned Owl, or "yellow-hammer" for Northern Flicker. Names such as "sparrow," "swallow," and "golden oriole" were generic and might have included several species. He also noted some as "three varieties of jay," or "humming bird (three varieties)." Other birds were "[Wild] turkey," and "large redbird with crest" or Northern Cardinal, but of the seventy-two birds listed, less than half can be specifically identified.

Some of these birds may have been unknown or, as was the case of the Mexican Jay, Irwin's specimen was the first to be collected north of the border (Baird and Ridgway 1873). Charles Bonaparte (1825), when considering the morphological subtleties of similar corvids, wrote in his description: "Amongst the numerous blue Jays and blue Magpies described by deferent authors, and magnificently figured of late, the pretensions of the present bird to novelty, will, at first glance, be doubted." Several early naturalists were "confounded" by the Mexican Jay, especially by the similarities between it and the Western Scrub-Jay.

In 1861, Irwin was involved in another historic event on the Butterfield Trail—the tragic "Bascom incident" near Apache Spring. A few days after the tragic episode began, Irwin was summoned on a two-day trip from Fort Buchanan to the vicinity of the "mail-station" where, he remarked "our presence disturbed a flock of buzzards [Turkey Vultures] some distance to the right of the trail leading to the chief's [Chiricahua Apache Cochise (ca. 1808–1874)] favorite camping ground, and, on riding over to the place where the birds had flown, the ghastly remains of six human bodies, upon which the vultures had been banqueting, were discovered" (Thompson 1935).

Irwin's presence at Apache Pass apparently had a profound effect on southwestern history, for it was the result of his actions that the atrocities continued against several Indians who had been taken prisoner. "It was I who suggested their summary execution, man for man. On [Lt. George N.] Bascom (1836–1862) expressing reluctance to resort to the extrement proposed, I urged my right to dispose of the three prisoners captured by me, after which he then acceded to the retaliatory proposition." Bascom then added three more hostages. Historian Dan L. Thrapp (1990) wrote that it may have been Irwin's insistence and uncompromising attitude, not that of the more "demurred" Bascom, that contributed to the hanging of six Apaches, thus precipitating more than two decades of intense conflict.

In 1894, Irwin received the nation's highest military award, the Medal of Honor, for his participation in the events immediately prior to his arrival at Apache Pass. He also achieved the rank of brigadier general the same year.

THE GENTLE NATURALIST, 1860–1870

🕉 DR. JAMES GRAHAM Cooper (1830–1902) was one of those fortunate individuals who was exposed to the wonders of natural history by an encouraging and supportive father. William Cooper (1798–1864), the father of James, was a founder of the New York Lyceum of Natural History at the age of nineteen and later gained acquaintance with such notable ornithologists as Thomas Nuttall, John J. Audubon, and Charles Bonaparte. The elder Cooper also contributed to the work of Bonaparte's *American Ornithology,* for which the author accordingly named the Cooper's Hawk *(Accipiter cooperii)* in his honor (Coan 1981). The Olive-sided Flycatcher *(Contopus cooperi)* was also named for him by Thomas Nuttall.

Spencer Baird first sent the younger Cooper on the railroad surveys of the Northwest, after which he served as a contract surgeon with the army in Oregon. But it is his work in the Southwest on which this brief biography of Cooper will focus. While working for the Geological Survey of California, a state agency, his first trip began with his arrival in Los Angeles in preparation for a journey to Fort Mojave.

In December 1860, Cooper (1869), in the company of an army wagon

Dr. James G. Cooper (1830–1902). (From Condor, *1902)*

supply train, began his 16-day "hasty journey" by mule to the remote outpost. His account as an experienced naturalist, although new to the Southwest, is an informative insight into the natural history of the region. He wrote that the coastal "Los Angeles Plains," except for the "little tree growth . . . along the streams," was "still brown and barren looking from the effects of the long dry season" but "in spring puts on a garb of the most beautiful green, varied with myriads of pretty flowers."

"Meadow Lark," Horned Lark, and American Pipit were "so numerous in places . . . to almost darken the air when they fly," and Mountain Plover ran "in scattered flocks over the driest tracts, or wheel in swift columns around the sportsman, their white under-parts sometimes shining like snow-flakes as they turn like their more aquatic cousins of the seashore."

The "Meadow Lark" Cooper observed was similar to that which Audubon described as the Western Meadowlark *(Sturnella neglecta)* in 1844. How-

Frontier Unrest

ever, its status, amid much confusion, shifted to that of a subspecies of the Eastern Meadowlark *(S. magna)*. But Baird still recognized their contrasting vocalizations. "In discussing the question of specific distinction between the two birds, the remarkable difference in their notes, as attested by all observers from Lewis and Clark down to the present day must be kept in mind" (Baird, Cassin, and Lawrence 1858). The song or "note" of the "larke" to which Captain Lewis alluded was that of *neglecta,* easily distinguished in the field by its glorious flutelike rendition (Cutright 1969). Despite the close resemblance of the two birds, the morphological and striking vocal differences were later reviewed and finally recognized to be sufficiently distinct to separate them into two species in 1910.

Climbing Cajon Pass through "extensive thickets of shrubbery, with occasional low trees," Cooper (1869) recorded a Red-tailed "Black" Hawk which was no doubt a dark-morph. On the high Mojave Desert, the road followed the Mojave River, which provides "a narrow tract of bottomland [that] forms a sort of oasis" for miles. Here he observed "Leconte's Mock-thrush" or Le Conte's Thrasher, and on reaching Soda Lake, the Black-throated "Finch" or Sparrow.

Located nearly on the 35th parallel, seventeen miles north of the conspicuously jagged Needles on the east bank of the Colorado River, Fort Mojave was established at Beale's Crossing because of an emigrant party massacre by the Mojave Indians in 1858. With the beginning of the Civil War in 1861, some army personnel transferred east and others resigned, as the North and South drew their battle lines.

Upon arriving at the isolated fort, Cooper wrote that the "valley of the Colorado . . . is ten miles in width, and formed of a succession of gravelly terraces, or mesas, with a narrow sandy bottom intervening, not over a mile wide. The whole upland has a most barren and desolate aspect. . . . The bottomland, however, supports a vigorous growth of cottonwoods, willows, and mesquite. . . . The river itself is so low in winter that the Indians can wade across with their heads above water."

After crossing the open desert and the "barrenness of the surrounding regions," Cooper, as a naturalist, "gazed with delight on the broad flashing stream, with its forest-clad banks, even though the trees were bare" in winter. During his six-month "faunal season" at the fort, his bird list lengthened to 119 species.

In January, Cooper noted "Malherbe's" or Gilded Flicker and the Tundra Swans that "appeared for a few days." As spring progressed he observed Bank

Swallows, Summer Tanagers, and Blue Grosbeaks along the river. Among the several nests he found were those of the Song Sparrow and the Yellow-breasted Chat. One nest of the chat contained three eggs of its owner and a single egg of a parasitic Brown-headed Cowbird. In summer, he observed the "strange Vulture-eagles" or Crested Caracaras along the river.

Cooper (1861) also discovered two new birds while at Fort Mojave. The first was a "beautiful little warbler" that he found by "its peculiar notes" and finally secured after "considerable watching and scrambling through the dense mesquite thickets." He wrote it is "undoubtedly a new bird, I have dedicated to the interesting little daughter of my kind friend, Prof. S. F. Baird." Lucy Hunter Baird (1848–1913), for whom Lucy's Warbler (*Vermivora luciae*) was named, was only thirteen at the time. In 1889, the Colima Warbler, the last of this genus to be discovered, was found in the Sierra Nevada de Colima, Mexico (Bangs 1925). It was forty years later that its occurrence was realized in the Chisos Mountains of the Big Bend region (Van Tyne 1929).

The second unique find by Cooper (1861) was "the smallest owl yet discovered within the United States," the Elf Owl, which he found "in a dense thicket, on a very windy morning." He named this tiny owl *Micrathene whitneyi* for Josiah Dwight Whitney (1819–1896), the eminent state geologist (1860–1874) in charge of the Geological Survey of California. Mount Whitney (14,495 feet), the highest peak in California, was also named for him.

Returning to the Pacific coast, Cooper continued with the survey, exploring much of California and its offshore islands. In 1862, he collected the first Zone-tailed Hawk north of the border near San Diego (Coues 1866b). Cooper, however, was mistaken when he labeled the specimen "Buteo harlani" (Grinnell 1909).

Cooper's major work on birds was *Ornithology* (1870), published by the Geological Survey. After editing the book, Spencer Baird expressed most favorably that it "is by far the most valuable contribution to the biography of American birds that has appeared since the time of Audubon" (Baird, Brewer, and Ridgway 1874). Following this, a collaborative effort by Whitney, Baird, and Joel A. Allen, of the Museum of Comparative Zoology at Harvard, brought about *The Water Birds of North America* (1884), published in two volumes. Although written by Baird, Brewer, and Ridgway, it was apparently Whitney, who, desirous to further science and incorporate Cooper's notes in the work, saw the project to completion "chiefly at his own risk and expense."

Cooper's contributions toward western avifauna were considerable, as indicated in a listing by Professor Joseph Grinnell (1902b). In 1893, the Cooper Ornithological Club was named in his honor, and the following year the *Nidiologist* became the club's first journal. Lasting only three years, this publication was followed in 1899 by the *Bulletin*. One year later the club's current journal took the name the *Condor* (Walsberg 1993).

The genial memory of Cooper was expressed by W. Otto Emerson (1856–1940), designer of the new publication cover and first President of the Club, when he wrote (1899): "he talked to us of Nature in all her varied forms; told of the birds, their songs, their flights, plumage and their homelife; of their loves and hates, joys and sorrows! . . . thus we saw Nature and her works through the eyes of one who loved and had long questioned and learned many of her secrets." Emerson (1902) later added that the strong attachment this gentle naturalist had for "Nature" was such that he "could never bear to see a tree cut down, or even have it used for fuel in his own home."

PURSUING THE ARIZONIAN ORNIS, 1863–1865

YOUNG ELLIOTT COUES (1842–1899) had the good fortune to grow up in Washington under the shadow of the great institution of collectibles, the Smithsonian, where his champion and close supporter, Spencer Baird, was assistant secretary. While only eighteen, at Columbian College (George Washington University), Coues, pronounced "Cows" (Cutright and Brodhead 1981), wrote his first paper in which he honored his mentor and lifelong friend. During this period his remarkable abilities as a systematic ornithologist and taxonomist were already materializing as he rummaged through the collections of the Smithsonian. It was here, as he was giving close examination to a series of shorebirds, that he discovered and described his first bird, Baird's Sandpiper (*Calidris bairdii*).

When naming this bird, Coues (1861) wrote: "In presenting to the scientific world this my *first* new species, I should do violence to my feelings, did I give it any other name than the one chosen. To SPENCER F. BAIRD, I dedicate it, as a slight testimonial of respect for scientific acquirements of the highest order, and in grateful remembrance of the unvarying kindness which has rendered my almost daily intercourse a source of so great pleasure, and of the friendly encouragement to which I shall ever feel indebted for what ever progress I may hereafter make in ornithology."

Elliott Coues (1842–1899).
(From Bird-Lore, *1902)*

In 1863, Coues, too young to receive an army commission, became an acting assistant surgeon. The following year he received his rank and was appointed assistant surgeon. He was then immediately transferred to the Southwest by his superiors at the request of Baird.

On June 15, Coues arrived at Los Pinos, approximately twenty miles south of Albuquerque, where he met Capt. Charles A. Curtis (1835–1907), who was acting quartermaster of a large supply train gathering in "rendezvous." Curtis (1902) took special interest in the young surgeon and recorded many events on the march "for fully five hundred miles through a hostile Indian region" to Fort Whipple located near Prescott in Arizona Territory. When he first spoke with the young doctor, Curtis recalled, Coues "was at that time still some months short of being twenty-two years old" and "had remarked, with pardonable pride, that he had been sent as surgeon in charge of our column at the request of the Smithsonian Institution, that he might 'shoot up the country between the Rio Grande and the Rio Colorado.'"

With preparations in reasonable good order, the large column began its slow trek west along the same route as the Sitgreaves and Whipple expeditions. With them were 300 cattle, 800 sheep, 92 wagons, 560 mules, 163 horses for cavalry escort and officers, and an additional company of infantry. It wasn't long before Curtis realized that Coues was a decidedly unusual sort of surgeon.

> *Ornithology was the Doctor's special cult. . . . For creeping, crawling and wriggling things he had brought along a five-gallon keg of alcohol. But the reptilian branch of his researches failed utterly in the early stage of the march, for the soldiers, in unloading and loading the wagon, had caught the scent of the preservative fluid, and, although it already contained a considerable number of snakes, lizards, horned toads, etc., the stuff, diluted from their canteens, did not prove objectionable to the chronic bibulants. Some of them, however, did look decidedly pale about the gills when the head of the empty keg was smashed in and the pickled contents exposed to view.*

The strange and sometimes daring activities of Coues en route brought considerable interest to the caravan as Curtis described:

> *Doctor Coues never ceased . . . making excursions along the flanks of the column. . . . Clad in a corduroy suit . . . he regularly rode out of column every morning astride of his buckskin-colored mule, which he had named Jenny Lind on account of her musical bray. Rarely did we see him again until we had been some hours in the following camp, but we sometimes heard the discharge of his double-barreled shotgun far off the line of march. He usually brought in all his pockets, and pouches filled with the trophies of his search, and when he sat upon the ground and proceeded to skin, stuff and label his specimens he was never without an interested group of officers and men about him. To any one interested to learn the art of preparing the specimens he became an earnest and painstaking instructor.*

Coues's assortment of tools, typical of all bird collectors, were scissors and various knives for dissecting and removing all internal parts but the skull and major bones, and cotton batting for stuffing the remaining cavity. Before the bird skin was stuffed and wrapped in paper for protection, arsenic was added as a preservative.

On one occasion during the march, with orders forbidding the discharge of firearms, Coues was reprimanded by his commanding officer, Capt. Allen L. Anderson (1836–1910), for shooting a desirable bird. His only excuse was, "I really could not allow this bird to escape without causing a serious loss to science." Anderson, enraged, lashed back, "Well, I shall deprive science of any further collections for a week by placing you in arrest and taking possession of your gun and ammunition" (Curtis 1902). The arrest, however, was suspended the following morning.

On July 2, after stopping at Fort Wingate for three days, the march continued to "the summit of Whipple's Pass . . . not far from the old site of Fort Wingate" where Coues (1878) discovered a new warbler. Later the same year he secured more specimens of the same bird in the vicinity of Fort Whipple. After receiving the specimens from Coues, Baird (1865) named the bird in honor of the young doctor's sister, Grace Darling Coues (1847–1925). Coues (1878) in expressing his deep devotion to her, later wrote: "Grace's Warbler [*Dendroica graciae*] is to me a bird of particular and not unpardonable interest, being the only species of this beautiful genus that it has fallen to my lot to discover, and bearing the name of one for whom my affection and respect keep pace with my appreciation of true loveliness of character."

Continuing his march to Fort Whipple on July 3, Coues (1865, 1866b) crossed the Continental Divide to Inscription Rock, "a huge mass of sandstone" where he "saw great numbers" of swifts. Baird (1854) had not recognized Woodhouse's (1853) naming of these as *Acanthylis saxatilis*. Instead, he accepted Kennerly's and Möllhausen's specimen and named it *Cypselius melanoleuous,* which was later revised. Following the nomenclature revision of the White-throated Swift by Baird, Coues (1866b) wrote: "I think there can be no doubt that the bird described by Prof. Baird . . . is the same as that briefly and somewhat incorrectly indicated by Dr. Woodhouse [(1853)]. . . . The chief discrepancy is the white rump mentioned by Dr. Woodhouse . . . from the imperfect observations he was only enabled to make, mistook the white patches on each side of the rump . . . [for] the large white cottony patches on the flanks [which] are long and loose enough to meet each other on the rump." Coues was the first naturalist to substantiate Woodhouse's visual description and at the exact same location.

On July 8, Coues described some of the misconceptions of the "delightful and equable climate" of the Southwest and the reality of physical sacrifices when exposed to the elements and living in the open with no shelter.

> *Last night we shivered under blankets, and blew our numb fingers this morning. By ten it was hot; at eleven, hotter; twelve, it was as hot as—it could be. The cold nights stiffen our bones, and the hot days blister our noses, croak our lips and bring our eye-balls to a stand-still. Today we have traversed a sandy desert; no water last night for our worn-out animals, and very little grass. The "sand-storms" are hard to bear, for the fine particles cut like ground glass; but want of water is hardest of all. For some time it has been a long day's march from one spring or pool to another; and occasionally more; and then the liquid we find is nauseating, charged with alkali, tepid, and so muddy that we cannot see the bottom of a tin cup through it. (Curtis 1902)*

While in the Southwest, Coues was posted at Fort Whipple, "a month's journey from anywhere." Originally located in Chino Valley, the garrison was moved a mile northeast of Prescott, the new territorial capital, immediately before Coues arrived. In a letter to Baird he wrote how his spirits were dampened by his lack of freedom and dramatic medical responsibilities. "The Apaches are so hostile and daring that considerable caution will have to tinge my collecting enthusiasm if I want to save my scalp" (Dall 1915).

Coues (1878) described "little troops" of "Plain" Titmouse (became Juniper Titmouse in 1997) on a nearby "scrubby hillside" and went on to explain that

> *[It] was a favorite resort of mine, not so much for what I expected to find there in the ornithological line, as for what I sincerely hoped not to find in the way of aborigines—for it was in full view of the fort, and much safer than the ravines on either side, where I have gone more than once to bring in the naked and still bleeding bodies of men killed by the Apaches. . . . the worst passions of both Red and White men were inflamed by atrocities exchanged in kind, and when practical ornithology in Arizona was a very precarious matter, always liable to sudden interruption, and altogether too spicy for comfort.*

It was in this vein that Coues's writings about birds were often "reminiscences" and his "whole notion of the lives of some of them—pervaded with local color."

Coues nearly lost his life during a "cruel massacre" of Hualapai Indians in the Juniper Mountains northwest of the fort. During the action the commander of the cavalry troop shouted a warning to Coues, who quickly turned to see a warrior, who appeared to him a "full ten feet tall. He was drawing his bow at me. I raised my gun, but it snapt, and it would have been all over with me, had not Captain [John] Thompson shot the Indian dead through the heart" (Cutright and Brodhead 1981). Coues, nevertheless, objected to some of the harsh army methods employed against the Indians and believed in their fair treatment by stating "the Apache has not broken faith with us oftener than we are proud to say we have with him" (Brodhead 1973).

During his first spring at Fort Whipple, Coues (1878) exhibited his keen competitive spirit when he mistook to his delight Lucy's Warbler as an undiscovered species, and then, when he realized its previous discovery, he exhibited his disappointment.

> Whilst rambling . . . along the little stream I heard a curious note . . .
> and was not long on the alert before I saw one of the modest vocalists,
> betrayed no less by the restlessness with which the bird skipped about in
> the budding foliage than by the singularity of its voice. Not recognizing
> the species, I made the usual sacrifice without delay, and was overjoyed
> to find, as I turned the dainty bird over and over in my hand, remov-
> ing every trace of blood and smoothing every ruffled feather, that
> I had taken a species new to me: for I had not then learned of
> Dr. Cooper's [(1861)] prize, and moments of discovery are always
> moments of pardonable enthusiasm.

Once, on an occasion of exceptional tenseness, Coues (1878) revealed a hint of his interest in mystical speculation about an unknown bird.

> I came to regard [Phainopepla] at last as great "medicine," so
> persistently did it elude me—now I could not get a shot at the shy
> thing—now a fair shot offered, but we had orders not to shoot for
> fear of discovery. It was a beautiful jet-black creature, showing a pair
> of white disks, one on each side, when it flew; generally seen amidst
> dense chaparral, dashing about with a nervous yet lightsome flight . . .
> now for a moment balancing with expanding wings and tail on some
> prominent spray, then darting into the air to secure a passing insect
> . . . once I listened to a superb piece of music which I am perfectly sure

*came from this mysterious stranger. It was growing dusk: the scene,
the camp of a scouting-party returning from unsuccessful pursuit of
some Indians . . . and men busy gathering for burial the charred and
dismembered body of a comrade, who had been killed and burned a
few days before on that very spot, where the wolves had afterward
fought for the remains. The bird of omen, for good or bad, appeared
in sombre cerements, and sang such a requiem as touched every heart;
the camp grew more quiet than usual, and we went to bed early.*

After collecting a pair of Plumbeous "Greenlets" or Vireos, which he
described in 1866, Coues (1878) gave a most touching account of

*the devotion birds so often show to their mates. The female, fatally
wounded by my shot, crouched upon a slender twig . . . breathing
heavily in dire distress. . . . Her mate in a few moments came flying to
her assistance. He alighted by her side, caressed her tenderly with his
beak, and seemed to beseech her, in low, sympathetic accents, to fly the
fatal spot. She gathered herself for the effort, but only fluttered fainting
to the ground . . . but her brave mate, heedless of my presence, never left
her side, nor ceased his fond attention, till he shared her fate.*

The Gray Vireo, "totally diverse from all others of North America," was
the second bird of that genus that Coues (1866b) discovered. However, he was
able to secure only one specimen, about which he added "the species must
be exceedingly rare, or I should have met with others." In his quest for "Arizo-
nian Ornis," Coues (1878) was unable to add anything about the habits of this
uncommon vireo other than it was not seen until ten years later when "Mr.
[Henry] Henshaw gave us the welcome contribution to its history."

Coues also collected a small "fulvous or buffy" empid that he introduced
as a new species because its appearance was "distinct, for the present, at least"
(1866b). This was done only after receiving a specimen for comparison from
"Mexico through the Maison Verreaux." Neither the "size or form" showed
any discrepancies, but the plumage coloration differed, along with their respective habitats. However, both birds were subsequently recognized as the Buff-
breasted Flycatcher, originally described by Giraud from among his "sixteen
new species" (Coues 1878). Although not accepted as new to Texas when submitted by Giraud, Coues's specimen was new to the United States.

The Black-chinned Sparrow was also first introduced to our fauna north of the border by Coues (1866b) while at Fort Whipple. Lt. Darius Couch first noted this lovely songster during the Mexican Boundary Survey near the border in the Mexican state of Coahuila in 1853. Spencer Baird also portrayed this sparrow in "Birds of the Boundary" (1859a).

Although Coues's (1878) acquaintance with the Loggerhead Shrike was limited to one taken at Fort Whipple, his eventual familiarity with it was exhibited by his long dissertation. He described it in part as "a bird of extraordinary spirit,—the stout, hooked beak, combining claw and tooth in one murderous instrument." A "gallant marauder," Coues continued, it "makes a rather imposing picture just then in his uniform of French gray with black and white facings, which fits him 'like a dream': the next instant—whish! he is gone, and the piteous cry of the Sparrow in yonder bush tell the rest of the story." Their prey are "devoured upon the spot, or carried to the 'cemetery' and stuck upon a thorn." Whenever Coues applied his pen, his remarks were an impressive display of technical knowledge and ostentatious diction.

While anxiously waiting for transfer orders east, Coues (1866a), at the "kind invitation from the Commanding-General of the Territory" took a September "pasear" down the Colorado River from Fort Mojave to the shady "117° Fahr." temperature of Fort Yuma. His steamboat cruise downriver was on the *Cocopah*, commanded by D. C. Robinson, the same captain who piloted the *Explorer* on Ives's expedition up the river in 1858.

In October 1865, Coues finally received his much desired orders to return to Washington. His route was from Fort Whipple west again to Fort Mojave, then across the Mojave Desert, past Soda Lake along the Mojave River, and over Cajon Pass to Drum Barracks on the southern California coast.

Coues eagerly looked up Dr. James G. Cooper, surgeon of the post, and they took short excursions together. To the young surgeon on his first trip west, Cooper was "an encyclopaedia of interesting biographies of the birds of the Pacific coast."

Sailing around San Pedro Bay, Coues was taken by the "unwatchful and unsuspicious" nature of the waterfowl. He took special note of Pacific Loons that

> swam as unconcernedly as tame Ducks. I remember once . . . leaning
> over the taffrail . . . with several of these Loons quietly disporting in
> the water below me, I felt that indeed I was a favoured ornithologist;
> for how many of my brethren have been able to study Loons in their

native wilds, only a few feet from the eye, and to note every motion, when in, or on, or over the water? Their appearance when returning to the surface after a long dive is peculiar; and really they look more like fish than birds. The bubbles of air clinging all over their backs gives them a beautiful spangled appearance; and when emerging from the sea, with a slight shiver of the feathers these spangles disappear; the water rolls away too, and the feathers are as dry as though they had not been submerged.

With a special interest in shore birds, Coues (1866a) was anxious to secure a specimen of the Snowy Plover along a sandy beach. With the thrill of uncertain excitement, he wrote "by racing at full speed through the heavy soft sand, joined to the exciting expectation of so soon seeing a new bird, I was quite breathless, and my heart was thumping furiously by the time I stepped on the moist sand. Yes! there they were sure enough, a flock of snow-white little beauties, dallying so fearlessly with the huge waves. I fancy my chagrin and disgust must have partaken a little of the sublime when, after blindly blazing away into the flock, I picked up a capful of—Sanderlings! Dr. Cooper's cachinations nowise tended to smooth my ruffled mental plumage."

In 1881, Coues, after an illustrious army career as a surgeon and scientist, was ordered once again, much to his frustration, to renamed Whipple Barracks, which ultimately brought about his resignation.

Coues progressed to be perhaps the most prominent naturalist and prolific writer of ornithology in the Southwest. Many naturalists benefitted from Coues as a knowledgeable and forceful editor. In this role, he would sometimes attach lengthy detailed comments for clarification or bring attention to any inaccuracies. His remarks about Heermann's (1853b) assumption that Northern Pygmy-Owl were active at "night" and that they are sometimes "caught perched on the branch of a tree, napping, during the day" serves as a case in point. Coues, having observed this owl "several times" at Fort Whipple correctly stated to the contrary, that it was "often found abroad in the daytime." Elaborating, he added: "The fact is, that it shows a decided tendency toward diurnal habits, and though thus . . . it will not likely be 'caught napping'" (Coues 1874a).

The 1870s were perhaps Coues's most productive years as a writer and editor. His erudite approach to the hundreds of articles he wrote, including the *Key to North American Birds* (1872) and the *Birds of the Colorado Valley* (1878) demonstrate his remarkable writing abilities. In the latter work, Coues also preformed a "gigantic task" in assembling a "Bibliographical Appendix" of North American Ornithology. Two installments followed (Coues 1879b and c). Leading English scientists of the period even "memorialized" him for this great work (Allen 1879).

It was also during this time that Coues was engaged with the Geographical and Geological Survey of the Territories under Ferdinand V. Hayden. Recognizing Coues's (1878) great literary abilities, Hayden wrote in the "Prefatory Note" that he made "natural history entertaining and attractive as well as instructive, with no loss of scientific precision."

Daniel G. Elliot (1835–1915), a founder of the American Ornithologists' Union like Coues, and who immediately preceded him as its second president (Coues was the third), wrote of his "brilliant mind" and his articulate treatment of the pen, which included classical Greek and Latin. "He possessed a command of the language gained by few and the beauty of his style and his felicity of expression has created numerous pen pictures of the habits and appearances of our wild creatures that have never been excelled by any writer, if indeed they have been equalled" (Elliot 1901).

Although Coues has contributed much to the enjoyment and understanding of birds, he deprived us of at least one additional work by an impulsive act he carried out only five years after his first departure from the Southwest. Unable to publish a three-thousand-page manuscript on Arizona birds, he read it to an individual uninformed on the subject, who passed an extremely unfavorable judgment. Distraught at the response to his labor of love, Coues abruptly burned the manuscript. A deep melancholy ensued whereby "a severe attack of illness in consequence" developed from which he eventually recovered, but not without considerable anguish (Elliot 1901).

RIFE IN THE CAMP, 1864–1875

AFTER ARRIVING FROM England in 1849, Edward Palmer (1831–1911) developed serious interests in botany and eventually became an untiring traveler and plant collector. On several occasions, however, he collected "all branches of natural history," which he sent to eastern individuals and institutions.

In 1864, Palmer began his experiences in the Southwest by briefly assisting Dr. James G. Cooper with the Geological Survey of California. After spending time in the east, he returned to the Southwest and was appointed acting assistant surgeon at Fort Whipple. Apparently Palmer was cordially received by Elliott Coues as the two men collaborated in their common pursuit of collecting specimens. Later, however, he learned to his chagrin that Coues, when taking their shared collections east, failed to add his name to botanical specimens they jointly collected. This apparent oversight or misunderstanding on the part of Coues was not conclusive, but it caused a bitter resentment of Palmer toward Coues (McVaugh 1956).

In 1866, Palmer was transferred to Camp Lincoln (renamed Camp Verde), on the Verde River at its confluence with Beaver Creek, forty miles east of Fort Whipple. While at this post, he contracted "intermittent fever" or malaria, a malady common to many areas in Arizona, which caused considerable difficulties to his health and ability to collect. Despite his limitations, he accompanied the troops on "7 foot scouts" against the Apache, during which he was able to collect. These actions were extremely physically demanding as they were "several days each, on foot in enemy country, each man carrying his pack for want of pack animals." It was on one of these "scouts" that he described a brutal massacre of Indians. "[T]he party resumed the march over every species of hill, valley, smooth and rough. . . . the command . . . assailed the foe, whom they found in caves. . . . Only two were seen to escape" (McVaugh 1956).

During the summer of 1866, Camp Lincoln was hardly functioning because malaria was "rife in the camp" with sixty men "down at a time." Palmer's situation deteriorated when, in addition to his fever, he received head injuries by being thrown from a mule. While being transported to the hospital at Fort Whipple, he was "so prostrated by fever" that he was unable to take along his collection. Although he received assurances from the post commander that his boxes would be sent on, it wasn't until a return visit three years later that he learned that "the things had been thrown away" (McVaugh 1956).

From Fort Whipple, Palmer transferred 155 miles southeast to Camp Grant at the junction of Arivaipa Creek with the San Pedro River. Malaria continued to give him dreadful problems when "for weeks at a time I was unable to do any collecting." In 1867, he was released by the army because of failing health, whereupon he returned east, but not before collecting 103 bird specimens that he sent to the Smithsonian Institution. Coues (1866b, 1868) first published a list of birds found in Palmer's "Prodrome" at Fort Whipple, which he then followed with another comparative list from Camp Grant that was supplied to him by Palmer.

The tall Palmer Agave, a name bestowed by botanist George Engelmann, honors this energetic collector. During June, July, and August, on rocky hillsides along the borderlands common to Arizona, New Mexico, and Mexico, these plants form a twenty-foot stalk. Branching, they bear dense clusters of golden-yellow floral tubes tinged in purple that attract numerous insects, nectar-gathering bats, and hummingbirds. Brilliant Rufous Hummingbirds, returning south along the Rocky Mountains, swarm to these agave and other available flowers. Of further interest, this spectacular hummingbird was discovered on the west side of Vancouver Island by explorer Capt. James Cook (1728–1779) and his men. His highly regarded young naturalist William Anderson (?–1778) died on the voyage from tuberculosis (Alden and Ifft 1943).

Palmer continued to collect on the Mexican mainland south to Guaymas, on Baja California, and on several of the adjacent islands including Guadalupe, approximately 220 miles southwest of San Diego, where he was the first naturalist to visit in 1875. His special interest on this island was to secure botanical specimens before the island was ravaged by introduced goats. Expecting to stay only six weeks, he was forced to remain for over three months with apparently some danger of starving before he was taken off by concerned friends from San Diego. Of the seventy-two specimens of island avifauna he secured, eight species were land birds.

Palmer sent the Guadalupe Island specimens to the National Museum, where Robert Ridgway (1876, 1877) published two papers on the collection, including a short discourse on their genesis and similarities with their continental allies. The significant ornithological discovery Palmer made on the island was the Guadalupe Caracara, a now extinct species, which differed primarily in plumage from the mainland and South American forms.

Palmer's notes on the "Calalie" or *Quelele* contain little about the natural history of this striking raptor except to describe the aggressive nature of it toward domesticated animals. Apparently Palmer noted they were quite common on Guadalupe (Ridgway 1876). "Hundreds of the birds have been destroyed by the inhabitants, both with poison and fire-arms, without noticeable diminution of their numbers."

The "Closing History of the Guadalupe Caracara" documents the demise of this interesting falcon (Abbott 1933). Known to the scientific world for only a quarter of a century, it was apparently tame in the manner of many island species and was easily shot or captured. By the early 1900s, the last specimens were collected into extinction. Unfortunately, few of those remain in museums.

THE INFANTRYMAN, 1866–1868

🐦 BORN AND EDUCATED in Switzerland, Johann Arnold Spring (1845–1924) joined the U.S. Army, received wounds in the Civil War, and was subsequently discharged in 1865. After recovering, he re-enlisted and was shipped to Drum Barracks near San Pedro, where he marched as an infantryman west to Fort Yuma, Camp Lowell, Fort Bowie, and Camp Wallen in Arizona Territory, totaling almost six hundred miles. After becoming first sergeant, he was discharged at Camp Lowell in 1868. He resided in southern Arizona the remainder of his life, turning to many pursuits including that of artist, writer, and teacher. The former Spring Junior High School in Tucson was named for this early educator in 1951.

Because of his interest in science, Spring sent a few specimens to the Smithsonian Institution, and is quoted in *John Spring's Arizona* (Gustafson 1966). Traveling between Camps Crittenden, forty-five miles southeast of Tucson, and nearby Wallen, which lies twenty-five miles further east, he encountered "several 'villages' of prairie dogs [extirpated about 1938]." Recalling their habits, he added: "They live in excavations surmounted by small tumuli full of holes that are used for ingress and exit by these pretty, graceful ground squirrels. . . . Their habitations, excavated by themselves, are shared with a small bluish-gray ground owl [Burrowing Owl], which lives with them in perfect harmony."

THE SOLDIER-ORNITHOLOGIST, 1871–1873

🐦 AS A YOUNG German emigrant, Charles Emil Bendire (1836–1897) enlisted as a private in the army, where he served with the First Dragoons in southern Arizona and New Mexico Territories during the 1850s. Unlike most great naturalists who matured in close association with the land and its creatures, Bendire's interest in birds developed while in the army in 1858 or 1859. A half century later, surgeon-naturalist Edgar A. Mearns (1907) noted that "more important" than the notes and specimens sent to the Smithsonian by Dr. Bernard J. D. Irwin "was his early training of Charles Emil Bendire . . . in exact methods of scientific observation. Bendire was then a young soldier of his command, attached to the hospital corps, and stationed at old Fort Buchanan and other camps in the vicinity of Fort Lowell and Tucson."

After the Civil War, Bendire traveled to his homeland. On his return, he was commissioned a cavalry officer, again serving at many posts throughout

Charles Emil Bendire (1836–1897). (Courtesy of the Smithsonian Institution Archives, Record Unit 95, Box 3)

the West. In 1871, First Lieutenant Bendire was transferred to Camp Lowell near Tucson where a group of local rancorous participants, together with a large number of Papagos (Tohono O'odham), trekked to Camp Grant, fifty miles to the northeast, to murder the sleeping Arivaipa band of Western Apache camping near Camp Grant.

Camp Lowell was situated in what is now the downtown section of present-day Tucson on the east bank of the Santa Cruz River. Rising in view are three prominent mountain ranges: the Santa Catalinas (9,157 feet) immediately to the northeast, the Rincons (8,482 feet) due east, and the Santa Ritas (9,453 feet) to the south. Fremont Cottonwoods and Mesquite lined the Santa Cruz River and a tributary, the "Rillitto" Creek. The surrounding flora, typically Sonoran Desert, included characteristic plants such as Creosote, Foothill Paloverde, several species of cholla, and on the rocky slopes, Ocotillo and Saguaro.

By 1872, Bendire (1892) was well underway in establishing his status as a great naturalist in a region that had been relatively unexplored by collectors. His encounters along "Rillitto" Creek with large "Black Hawks" were, at that time, as confusing to him as they are for many observers today. The identity of two species of these hawks remained a quandary for him, until he was able to inspect several specimens and correspond with other naturalists, including Dr. Edgar A. Mearns and Robert Ridgway. Choosing not to shoot a pair of hawks that "kept circling around and over" him caused for him some considerable difficulty in determining their species. He at first "wrongly identified" them as Swainson's Hawks but was later able to sort out this pair and correct his identification to that of "Mexican Black Hawks" or Common Black-Hawks.

Another experience along "Rillitto" Creek with "Black Hawks" created an indelible memory for Bendire. After a second thrilling visit to a Zone-tailed Hawk nest forty feet high in a cottonwood tree, he wrote: "Climbing to the nest I found another egg, and at the same instant saw from my elevated position something else which could not have been observed from the ground, namely, several Apache Indians crouched down on the side of a little cañon. . . . They were evidently watching me, their heads being raised just to a level with the top of the cañon."

Continuing, Bendire wrote:

In those days Apaché Indians were not the most desirable neighbors, especially when one was up a tree and unarmed; I therefore descended as leisurely as possible, knowing that if I showed any especial haste in getting down they would suspect me of having seen them; the egg I had placed in my mouth as the quickest and safest way that I could think of to dispose of it—and rather an uncomfortably large mouthful it was, too—nevertheless I reached the ground safely, and, with my horse and shotgun, lost no time in getting to high and open ground. . . . I found it no easy matter to remove the egg from my mouth without injury, but I finally succeeded, though my jaws ached for some time afterward.

Bendire must have been a man of unusual nerve for he returned "within an hour" in an attempt to secure one of the adult hawks.

His dauntless actions were noted by C. Hart Merriam (1897), when he stated that "Bendire was a man of energy, perseverance and courage, and in our Indian wars naturally took a prominent part. This part was sometimes that of a dreaded foe . . . sometimes that of a peace-maker, as when in the midst of the bloody Apache war he boldly visited the camp of Cochise, the celebrated Apache chief, and induced him to abandon the war path."

Of the several brown thrashers in the Southwest, two appear very similar except one has a long curved bill, while the other has a much shorter straight bill. To all collectors prior to 1872, these distinctions were overlooked or thought to be the more common Curved-billed Thrasher. Bendire, however, recognized these and other differing aspects, including their eggs, nests, and habits. It was only through his insistence and after a series of specimen and correspondence exchanges with Elliott Coues, and Coues with Robert Ridgway, that Coues finally described and named the bird "Bendire's Mocking-thrush" or Bendire's Thrasher (*Toxostoma bendirei*) in 1873.

The Rufous-winged Sparrow, the last discovery in that genus, was another new bird credited to Bendire while at Camp Lowell. This sparrow occurs locally in uniform terrain where there is ample grass, cholla, and mesquite thickets.

Bendire (1892) also added three borderland birds to our avifauna north of the border while exploring the Tucson area. The first was the Gray Hawk, which he considered "one of the handsomest Hawks we have; graceful and quick in all its movements." Another was the lovely twitching Painted Redstart previously noted on the south side of the lower Rio Grande border in the *Pacific Railroad* and *Mexican Boundary Survey Reports* (Baird, Cassin, and Lawrence 1858; Baird 1859a).

Regarding the third bird, Bendire (1892) admitted his shortcomings when he added several specimens of the Ferruginous Pygmy-Owl taken along "Rillitto" Creek. "Unfortunately, I was not then an adept in taxidermy; the skins made by me in those days looked as if they had passed through the jaws of a hungry coyote, and they were only useful in determining species." Now considered extremely rare, this aggressive diurnal owl was a "common occurrence" among the "cottonwood that fringes the Gila and Salt rivers of Arizona," according to George F. Breninger (1898).

In 1873, Bendire, after eighteen months at Camp Lowell, was promoted to captain and then transferred. In the same year, the post was relocated seven miles northeast of Tucson to "Rillitto" Creek and designated Fort Lowell. Outside of Bendire's military obligations as an officer, which he took very seriously, his "principal object was to study the nesting habits of our birds, as well as to collect their eggs." He was also the first to discover the nests of Lucy's Warbler and Spotted Owl (Coues 1878; Bendire 1892).

In 1883, Bendire took an eleven-month leave of absence during which, at the request of Spencer Baird, he was given the title of honorary curator of the Department of Oology at the U.S. National Museum. During this period he applied himself with great pride to the reworking of the huge collection of eggs, to which he contributed approximately eight thousand.

Bendire (1891) summed up his deep concern regarding egg collectors, a popular hobby of the day, when he wrote: "Unless the would-be collector intends to make an especial study of oölogy and has a higher aim . . . he had better not begin at all, but leave the nests and eggs of our birds alone and undisturbed. They already have too many enemies to contend with. . . . His principal aim should be to make careful observations. . . . This period comprises the most interesting and instructive part of the life history of our birds."

Most ornithologists who knew Bendire referred to him with the highest of accolades. However, Coues (1897), in a pretentious mood, asserted that Bendire was "my original discovery," and that he led "the bumptious and captious German soldier" to a "consummation with which we are all familiar." His blatant remark referred to the congenial friendship and cooperation which developed between Baird and Bendire, enabling Bendire, in later years, to ultimately achieve his prominent position in the field of ornithology.

Bendire, after Baird's passing but with his prior encouragement, wrote the splendid two-volume *Life Histories of North American Birds* (1892, 1895) recalling his experiences, along with the observations of numerous other contributors who felt honored to be included. He was also one of the founders of the American Ornithologists' Union.

Dr. James C. Merrill (1898), a distinguished army surgeon and naturalist, wrote of Bendire's character: "Frank yet reserved, bluff, honest and truthful to bluntness, he had the courage of his convictions, which he did not fail to make clear when occasion required. Simple in habits, unselfish, and always ready to help others."

6 WEST OF THE 95TH MERIDIAN

"Of Mexican birds, that extend across our lines, and find their northern
limits within our areas, there, doubtless, yet remain quite a number
to be discovered, and these not mere stragglers, but such as exist in
considerable numbers. These will probably be found principally in the
southeast [Madrean Archipelago], as there the mountains, continuing in
an unbroken range from the table-lands of Mexico, afford a highway, as
already ascertained, for quite a number of otherwise extralimital forms,
which will be still further swelled by additional research."
—HENRY W. HENSHAW (1875a)

DURING THE LAST three decades of the nineteenth century, the exploration of the West continued with an emphasis on mapping and economic
resources, especially geology and mineral development. Other scientific disciplines, however, were included in the surveys, enabling them to be more comprehensive. Emphasis was now placed on the interior rather than prospective
roadways or boundaries. In 1879, an act of Congress abolished all existing field
surveys and created the U.S. Geological Survey within the Department of the
Interior under Clarence King. The surrender of Geronimo and his band in
1886 greatly reduced the active participation of the army as a fighting unit in
the Southwest. This also brought an end to the select cadre of naturalists who
served as officers, enlisted men, or under contract with the army.

THE WHEELER SURVEY, 1873–1884

☙☙ FOR A DECADE following 1869, the Geographic and Geological Explorations and Surveys west of the 100th meridian encompassed a huge land mass, which included the Mexican border region of Arizona, New Mexico and southern California, on which this narrative will focus. The duration of this and other government surveys depended on yearly appropriation bills that had to be approved by Congress. The key members of the field surveys, therefore, retired to Washington at the end of each season to write progress reports and await in anticipation their future for the coming year.

It was probably through the acquaintance of Thomas M. Brewer that Henry Wetherbee Henshaw (1850–1930), "a promising young bird collector," was recommended to Spencer Baird to be a member of the Wheeler Survey. After his appointment, Henshaw reported to Lt. George M. Wheeler (1842–1905), a U.S. Military Academy graduate and topographic engineer in charge of the survey. It was this officer, Henshaw (1919) noted, "who, to no small extent, was to be master of my fate for the next decade." Mountain peaks in Nevada and New Mexico, and the narrow-leaved Sotol (*Dasylirion wheeleri*) are named for Wheeler.

Through his "Autobiographical Notes," Henshaw (1919, 1920a) provided a rich backdrop for his early field experiences with this expedition, which was entering a relatively unknown frontier. Regarding the survey priorities, he was quick to note that the "collection of natural history specimens formed a very small part" and "was wholly subordinate" to the geographic and cartographic endeavors. Each field party, of which there were from five to ten, usually consisted of a young Military Academy officer in charge of at least one topographer, an odometer assistant, a cook, several packers, and, when practical, a geologist and naturalist.

In addressing the circumstances of his labors, Henshaw expressed his synergetic relationships with various members of his party, including his reliance on the packers. Among them were a few typical "bad men of the west," but "each and all of them had not a few redeeming traits" from which he "received many kindnesses at their hands." Their job each morning was to round up the mules—which sometimes involved backtracking "to the last camp, often fifteen or twenty miles away." Their daily routine was to secure at least three hundred pounds of boxes and bundles of "camp impedimenta" by the use of the "diamond hitch" to the "aparejos," or padded saddles, on sometimes twenty "unwilling" mules. Occasionally, on the rough mountain trails, an animal would be upset and roll helplessly down the slope into a stream, which required the packer to "plunge into the water, rescue the drowning animal and its pack, and replace the load." The packers or, as Henshaw called them, "comrades of the trail," were "very communicative and entertaining" even in the most trying of conditions. "Polished manners and scholarly attainments were foreign to the packer's calling, and usually they lacked the gift of eloquent speech, but when things went wrong with the train, and stubborn mules needed rebuke, their outbursts of profane imaginings amounted to real eloquence."

After several thousand miles on the trail, Henshaw's thoughts turned to another dependable companion, the "American Mule," which he held in great admiration, perhaps bordering reverence.

> *Overlooking the bar sinister which attaches to his birth, and judging him in a friendly and not a hostile spirit, his native virtues are many, his faults few, and those chiefly due to bad treatment. Stripes and blows he never forgets and rarely forgives, and, as he has a good memory, he sometimes waits long for an opportunity to retaliate. That he is stubborn cannot be denied, but he is also patient and long suffering, and such is his endurance that he survives and even prospers under circumstances, such as lack of food and water, which would quickly prove fatal to . . . the horse. Strong and enduring, he is a tower of strength on the dizzy mountain trail and in the parched desert. He soon comes to recognize the kind touch and decent treatment of the master, and to kindness and consideration he quickly responds. A good riding mule . . . is the easiest riding animal in the world. . . . I raise my voice in praise of this much misunderstood and much underestimated animal.*

Henshaw's opinions of camp cooks also reflected his highest regard for the necessity and importance of these dedicated men. "To cook a savory meal . . . in the open air, on a windy day, often under a rainy sky, with bits of brushwood for fuel, and a horde of hungry men demanding something to eat, is given only to the elect and to few of them. I hold in grateful memory of such cooks whose triumphs over the trials and tribulations that beset their calling entitle them to golden crowns."

Life in the open, in which everybody slept on a "luxurious couch" such as only the ground could afford and with a saddle as a pillow, suited Henshaw

Henry Henshaw (1850–1930) and his dependable trail companion, the "American Mule"; Camp Bidwell, California, 1878. (Courtesy of the Smithsonian Institution Archives, Record Unit 7150, neg. 2000-7971, AOU 1873–1977)

while in the field. Their daily task began with an early breakfast, and they would be in the saddle by six or before. The cook followed a pre-arranged route until four or five in the afternoon, when a camp would be selected near water, wood, and grass if available. The U.S. Commissary was sufficient "to keep the men in abounding good health," but the "naturalist, with a trusty shotgun . . . was always a welcome addition to any party."

Astride his mule, Henshaw's personal outfit consisted of rough clothing and, attached "somewhere" about his person, an insect net. A double-barreled shotgun was "slung on the horn of the saddle" which, incidentally, was "a fearsome weapon to use on hummers and other small species." His general appearance, "quite out of the ordinary," suggested that of "an escaped lunatic or a highwayman." Occasionally, when he "chanced to meet a solitary horseman on the lonesome trail [he] saw him slip a hand furtively behind to make sure that his gun was ready." In meeting a concerned or tentative stranger, he added, "The insect net particularly excited curiosity, but when I explained I was a 'bug hunter from the Smithsonian' I was at once accepted as harmless."

Besides being provided for the chief of each party, tents were also generally available for the naturalists, who worked into the "wee small hours" preparing specimens they collected during the day. It was not uncommon to prepare specimens considered rare by immediately dismounting to use the only dissecting table available—the saddle of a mule. This procedure, under the best of circumstances, often tried the patience and skill of the collector, especially with bothersome flies and a "restive" mule.

After preparing specimens, properly drying them was a continual problem, as they seldom camped "two nights in the same place." Henshaw's method was to place them in paper cones and tack them to the bottoms of trays that fit into a stout box. "Providing all went well, the skins dried in very good shape, but the stampede of the pack animals, no very rare event, was likely to deposit the skins in a heap in one corner of the tray, a catastrophe which necessitated much labor in reshaping them."

In June 1873, Henshaw, before moving south, began his field season along the Rio Grande near Fort Garland in southern Colorado, where he experienced a humorous, but frightful collecting incident. While among some dense thickets, he shot a small bird, when, "following the report, a large bear tore through the brush only a few feet away. That he made excellent time his tracks subsequently showed, but as I had the down hill side of the proposition I am sure he did not run so fast as I did. On returning to claim my specimen, I found that the bird had actually fallen on his bearship as he lay snugly curled up asleep."

Henshaw's first trip into Arizona and New Mexico Territories began in July along their common northern boundary (Henshaw 1875b). Preceding Henshaw by two years, Dr. Ferdinand Bischoff, a German collector for the Wheeler Survey "wandered off into [this] desert and was never heard of again" (Dall 1915). Leaving Fort Wingate, Henshaw, in the company of Dr. C. G. Newberry, an assistant surgeon, traveled slowly to Inscription Rock where he, like some of the earlier naturalists, remarked about Samuel Woodhouse's (1853) description of the White-throated Swift. "To understand the mistake made by Dr. Woodhouse in describing the bird as possessed of a white rump, one need only place himself at an elevation where he can look down upon the bird as it courses the air below, or nearly on a level with himself, when he will instantly perceive what appears to be a continuous white patch across the rump, but which in reality is merely due to the apparent overlapping of the conspicuous white flank-tufts."

While visiting the Zuni Indians, Henshaw (1875b) noted their use of the Bald Eagle for ceremonial dress. The Indians kept the helpless birds penned in "wicker-inclosures" where "they presented a most lamentable appearance, as their bodies were devoid of feathers, which had been plucked out long before." He also observed wild Bald Eagles, including a pair in a canyon a few miles south of Fort Apache. Commenting on the cosmopolitan "Fish Hawk" or Osprey, Henshaw observed it occasionally "busily employed in its vocation on the small streams. On the Gila, however, which is plentifully stocked with fish, it seems to find a congenial home, and is quite abundant along its banks."

While at Fort Apache in the White Mountains, among the higher coniferous forests of pine and spruce, Henshaw noted the isolated southernmost population of the bold Gray Jay. He also took note of the "short simple melody, but beautiful from its extreme sweetness of expression" given by the Western Tanager.

Traveling south, Henshaw (1874) crossed the Gila River to newly relocated Camp Grant at the southern base of the Graham (Pinaleno) Mountains. Encountering a large "Refulgent" or Magnificent Hummingbird, he stated, it has "for the first time . . . been ascertained to inhabit the United States." After reaching Fort Bowie, Henshaw (1873) returned by way of the headwaters of the Gila River, where he found Wild Turkeys "literally swarming," before completing his return circuit to Fort Wingate.

The following year in 1874, Henshaw was accompanied by Dr. Joseph T. Rothrock (1839–1922) and James M. Rutter, in addition to a small escort of cavalrymen, on another trek, which extended further south into Apachería. Edward W. Nelson (1932), a long-time friend and associate, recovered a series of letters written by Henshaw to Ruthven Deane and C. Hart Merriam, which disclosed many interesting facts about that summer and fall season. In these letters, Henshaw described his Fort Apache neighbors in rather cautious terms. "The Apaches swarm about our camp which is on the bank of the White River near the Fort. One or more always accompanying me collecting and, as they have the eyes of a hawk, are as good as a retriever. Walking along I often hear a low 'coosh-coosh' and, turning, had one of them pointing at a bird in a tree top that I can hardly see. They seem to think it is the greatest sport in the world. . . . Apaches may take a liking for my scalp which, by the way, I have had shaved clean to lessen its market value." Henshaw went on to admit that the Indian's eyesight was greatly superior to his by acknowledging that he "earned their well merited contempt for the white man's blindness."

Before leaving Fort Apache, Henshaw (1875b) added a new bird north of the border, the showy Red-faced Warbler. This warbler was one of the "sixteen new species" of birds claimed by Jacob P. Giraud as having been collected in Texas. Henshaw also encountered this lovely warbler eighty miles south on Mount Graham, named for Col. James D. Graham, who served on the Mexican Boundary Survey (Williams 1987).

Henshaw, returning the second time to Fort Grant, viewed with great interest lofty Mount Graham (10,713 feet) immediately north of the post. Up to this time, only naturalists Col. George McCall, Adolphus Heermann, C.B.R. Kennerly, Arthur Schott, Dr. B.J.D. Irwin, Dr. Edward Palmer, and Charles Bendire had traversed portions of this mountainous region in search of birds. Their work demands and prudent safety limitations confined them to broad valley grasslands and certain low mountain passes. They seldom, if ever, ventured very far up the canyons into the unexplored higher mountains. Henshaw, however, on this trip to Arizona Territory, was the first naturalist to investigate the upper realm of "pineries" and climb still higher to at least ten thousand feet on Mount Graham. On the upper slopes of this mountain, the highest north of the border lying within the Madrean Archipelago, he entered a mountain island of tall Engelmann Spruce forests.

Rising above the Mexican highlands in southeastern Arizona and southwestern New Mexico are several high mountains that have a floristically distinct Madrean Woodland, which extends north across the border from the Sierra Madre in Mexico. These isolated ranges form the northern limits of the larger Madrean Archipelago and include five over nine thousand feet, two over eight thousand feet, and five over seven thousand feet in elevation. The valleys separating these mountains are near four thousand feet in the east, while those to the west near Tucson are approximately twenty-four hundred feet.

This biotic region also delineates the northward movement of several bird species, including the mountain dwelling "Mexican Snowbird" or Yellow-eyed Junco, which is found on some of these mountains (Marshall 1957). It was from Mount Graham that Henshaw (1875b) added this junco to the avifauna of the United States. On August 1, 1874, he secured the nest of the Magnificent Hummingbird, "saddled on the horizontal limb of an alder, about twenty feet above the bed of a running mountain stream, in a glen which was overarched and shadowed by several huge spruces, making it one of the most shady and retired little nooks that could be imagined."

Henshaw was very familiar with the habits of Cassin's Sparrow and told of the "indescribable sweetness and pathos" of their distinctive song, which he heard while on a night march from Camp Grant to Camp Bowie: "I do not think an interval of five minutes passed unbroken by the song of one of these sparrows. Ere fairly out of hearing of the notes of one performer, the same plaintive strain was taken up by another invisible musician a little further on, and so it continued till just before dawn."

Traveling south to Camp Bowie during the same year that Apache Chief Cochise died gave Henshaw clearly some reason for apprehension when he remarked, "what effect his death will have none can tell" (Nelson 1932). Nine miles southeast of Camp Bowie, a highly visible rhyolite monolith was later given the name Cochise Head, because of its resemblance to the famed chieftain's profile. Thirty-five miles southwest and across the Sulphur Springs Valley in the Dragoon Mountains is another massive granitic rock formation (Cochise Stronghold) that sometimes served as his home and fortress.

While near Camp Bowie, Henshaw (1875b) obtained a single hummingbird specimen whose identity eluded him until after his report was completed. Describing his encounter with this trochilid feeding on Palmer Agave, he wrote that "their spreading bunches of blossoms, dotted the rocky hillsides in every direction, and gave a strange, weird aspect to the landscape. Around these the hummingbirds congregated, showing an especial liking for the nectar of the flowers, or possibly finding in them rich storehouses of the minute forms of insect life."

Sending his specimens east ahead of his arrival, he waited in excited anticipation for the identities of those unknown to him. Anxiously, Henshaw wrote Merriam: "Let me see, haven't I found some new things among my birds which have come in since I wrote home? Certainly I have." One specimen discovered by William Bullock in Mexico proved to be a Lucifer Hummingbird after examination by G. N. Lawrence (1877). Upon learning this bit of news, Henshaw added in jocose humor, "Father Baird and the rest have always been foretelling [it] would some day turn up within our realm" (Nelson 1932).

Some of Henshaw's (1875b) descriptions about where certain species were collected appeared to be in conflict between Fort Bowie and his next destination, Camp Crittenden. Those localities in question, however, have been resolved to reflect his later notes (1920a). Leaving Fort Bowie, his route led to the southwest approximately eighty-five miles, probably through Middlemarch Pass, a low gap in the Dragoon Mountains, across the San Pedro River to Camp Crittenden. Half a mile southwest along Sonoita Creek lay the ruins of old Fort Buchanan, while the forested Santa Rita Mountains (9,453 feet) rise steeply less than nine miles northwest. Close to the Mexican border, and twenty-five miles southeast across several ridges, an equally high range looms, the Huachuca Mountains (9,466 feet).

Although Golden Eagles were seldom observed in the Southwest by early naturalists, they were no doubt fairly common in prey-rich areas with suitable nesting sites. To fill this rather significant void, we will deviate more than a century to the spring of 1983 to note a personal observation seven miles east of old Camp Crittenden. With the rugged Whetstone Mountains as a backdrop, an adult eagle took off from an earthen stock tank and flew leisurely east, while overhead, a half dozen Chihuahuan Ravens gathered and began circling. Gaining considerable height, the ravens converged with a dash on the ever-alert eagle, whereupon the raptor suddenly rolled on its back and, in an instant, thrust upward its sharp talons and clasped one of its unsuspecting tormentors. Taking it to the ground, the eagle disappeared beneath a lone mesquite about seventy-five yards from the road on which I was traveling. Repeated swoops were made over the bush by the ravens, but they soon lost interest. Then, after a considerable period, feathers commenced flying from the mesquite as the eagle began a rapid process of plucking the hapless raven. Finally, after a two-hour wait, the eagle departed on heavy wings with its featherless prey, leaving only a large pile of black fluff.

Henshaw's observations and collections were, nonetheless, exciting near Camp Crittenden. Although Botteri's Sparrow *(Aimophila botterii)* was taken earlier near the border at Nogales, Sonora, by C.B.R. Kennerly, Henshaw (1875b) was the first to add it north of the border. Philip L. Sclater (1857b) named this obligate of tall Wright Sacaton and salt grass prairies from a specimen collected in Mexico by Signor Matteo Botteri (1808–1877), a teacher who had relocated from Dalmatia on the Adriatic coast.

Among the oaks in the foothills, Henshaw secured another Mexican bird new to the United States—the only "smoky-brown"–backed woodpecker included among our fauna. It had been originally collected in Mexico by T. G. Mann, who sent it to Hugh Edwin Strickland (1811–1853), an English scientist interested in developing uniform international standards on zoological nomenclature. In 1845, Strickland showed the specimen to the French ornithologist Alfred Malherbe, who named it Strickland's Woodpecker *(Picoides stricklandi)*.

Henshaw (1920a) was also "greeted" by the high pitched, "shrill notes" of the vociferous Sulphur-bellied Flycatcher, another addition to birds north of the border. He also observed "considerable numbers" of "Nonpareil" or Painted Bunting in the vicinity of Camp Crittenden and at Camp Bowie (1875b). From one of the southeast canyons of the Santa Rita Mountains, Henshaw added another bird from south of the border, the striking Broad-billed Hummingbird, an earlier discovery by Bullock while he was in Mexico.

Near the post among willows "growing in small clumps or fringing the streams," Henshaw (1875a and b) observed the extremely rare Willow Flycatcher (*Empidonax traillii*) and recorded that their occurrence was "more or less abundantly throughout the Territory." This flycatcher was named for Scottish Dr. Thomas Stewart Traill (1781–1862) by J. J. Audubon in 1828 in gratitude for kindness he received from him while in Britain.

Henshaw resumed his trip, passing by way of "Sienega" and on to the garrison at Tucson. "I found myself at [Fort] Lowell turned loose among [Rufous-winged Sparrows] of Bendire fame and though I found the little beggars exceedingly difficult to capture I felt tolerably well satisfied when I counted up to 19." Although his stay was only four days, he added "I believe [it] to be one of the best points for collecting I ever saw. To birds there is no limit, nor do I believe it is by any means exhausted yet" (Nelson 1932).

Mid-September of 1874 found Henshaw (1875b) again high in the coniferous forest of the Graham Mountains where, among the "hundreds of feathered migrants," he collected the Olive Warbler, another addition north of the border. It, too, was a member of the "sixteen new species" collected by Giraud and, like many of the others, it was challenged as to its occurrence in the United States before Henshaw noted it.

Returning to Camp Apache in October, Henshaw revealed one of the major discomforts that plagued him during much of that season. "We got in the southern part of Arizona into a malarial region and as the season was a most unusually sickly one, so sayeth the eldest inhabitant, we have all suffered. Myself with a fever which played the d—l with work keeping me on my back much of the time, and remainder not much better" (Nelson 1932).

Returning to Washington in a happier vein, Henshaw wrote to C. Hart Merriam: "All the Ornithological Dignitaries were taken completely by surprise by the showing of tropical birds and it looks as though it might necessitate some change in faunal limits. So, my boy, I suppose I am to say that the season's work has been a fair success. Coues is as cheerful as ever. Brought home 600 [bird] skins besides many large mammals."

Although Henshaw wrote (1875b) "The Ornithological Collections" of the Wheeler Survey, Henry C. Yarrow (1840–1929), working primarily in the northern sector, was acting assistant surgeon and senior field naturalist for

ASTURINA NITIDA VAR PLAGIATA

Gray Hawk, Asturina
nitida *(Latham), 1790.
Painted by Robert Ridgway
(ca. 1874). (Henshaw 1875b)*

the expedition. Yarrow's Spiny Lizard was named for him. The striking border-land bird plates from Wheeler's expedition painted by Robert Ridgway display his remarkable artistic skill, in addition to his skill as a renowned systematist. Although William Brewster's (1876) comment criticizing their "point of coloration" may be essentially correct for some of them, others remain among the finest portraits painted thus far. Although the birds were not placed in natural landscapes, Ridgway's approach achieved a lifelike character, especially through accurate perching postures with sparkling eye reflections.

In 1875, Henshaw (1876) continued his work in southern California, including a visit to Santa Cruz Island. From that island a decade later, he described the Island Scrub-Jay, which was finally recognized by the American Ornithologists' Union as a distinct species in 1995.

Henshaw also traveled to old Fort Tejon, but found California disappointing compared to his previous two years in Arizona Territory. Apparently the vegetation was by no means what he had been led to believe by certain newspaper accounts: "[T]he regions traversed by our party has been one of the most desolate I ever saw, a desolation none the less dreary to look upon in that it comes not directly from nature but from the agency of man. A very large proportion of the country at large lying outside of such large towns as Los Angeles and Santa Barbara has been and is so completely overstocked with cattle and sheep that had fire swept over the country it could hardly have left it more bare. The hills for miles and miles have been completely denuded of grass and to a great extent of shrubbery too, for hungry stock are by no means gourmands" (Nelson 1932).

The works of man were beginning to extract a toll on birds in several ways. When Henshaw (1920a) was unable to fulfill his desire to see a "noble" California Condor, he remarked: "No man's hand was raised against it, but hundreds fell victims of the poisoned meat that the sheep herders put out" indiscriminately for predators. In 1884, while on a return trip to San Antonio Mission in the area of Monterey, he "saw four individuals circling about high in air and a notable sight they were."

In 1877, Charles A. Allen (1841–1930), an amateur collector from central California, sent to William Brewster a trochilid that was similar in most respects to the Rufous Hummingbird except that it had a greenish back. Henshaw (1877), on his return east, was attracted by Brewster's specimen and, thinking it to be a new species, he proceeded to describe and name it for Allen. However, because French ornithologist René Primevère Lesson had described another specimen taken from the same area in 1829 and applied its scientific name, only the common name of Allen's Hummingbird has been retained.

Declining the presidency of the American Ornithologists' Union, Henshaw served as its vice president from 1891 to 1894 and from 1911 to 1918. He also became the second chief of the U.S. Biological Survey from 1910 to 1916.

At the age of seventy, Henshaw wrote of his maturing attitude toward taking life: "for a number of years I have found it impossible to kill birds, or, indeed, to take the life of any living creature. I believe that sentiments akin to these are rather common among . . . [those] no longer young . . . one is better able to appreciate the value and significance of life, of whatsoever form, and to desire to cherish rather than destroy it" (1920a).

Edward Nelson (1932), a great admirer of Henshaw, remarked: "He was an acute and sympathetic observer of nature and her ways and loved all living things and their habits and relations to their surroundings. The exquisite beauty of form and color so lavishly displayed among birds and some other forms of life especially appealed to him."

A PLUCKY CHAP, 1874–1876

Like most naturalists of the period, Charles Edward Howard Aiken (1850–1936) was an easterner; however, as a young man seeking new horizons, he pioneered west to reside in Colorado. In 1874, three years after his arrival, he collected for the Wheeler Survey mainly in the Rocky Mountains. In 1876, the twenty-six-year-old naturalist made a five-month trip from Colorado Springs south into New Mexico Territory crossing the Rio Norte, west past old Fort Wingate into Arizona Territory, and then south to Fort Apache. Continuing to the Black River, he made two trips to Camp Goodwin on the Gila River before returning to Colorado.

In writing a biography of Henshaw, Edward Nelson (1932) retrieved a letter sent to Ruthven Deane by Henshaw that reveals his fondness and jovial envy for Aiken in his dangerous jaunt. "Aiken writes me from way down in New Mexico. He had had up to that time but little success and something like a hard time. His mules had but two drinks in three days. Was accompanied by only a small boy [fourteen-year-old Ed Rice] as assistant. . . . Fear he will not meet with the haul I anticipated. He may, however, strike it rich down in Arizona if he don't meet with hostile Apaches. Then Allah preserve him. He is a plucky chap. Knowing that country as I do I wish I were with him."

Aiken's journey was mainly by a mule-drawn wagon, which carried "his outfit" and led his saddle pony (Warren 1936). The Indians were apparently "quiet" that summer, however, one did steal his dog. Arriving at Camp Goodwin on his second trip, he developed "an attack of fever" whereupon he presented a "letter from Prof. Baird to the commanding officer. They seemed to distrust me at first because of my hard appearance, but after a little conversation it was all right." Under the care of the post doctor, he regained health and also recovered his dog Beech.

Aiken's (1937) observations of fifty bird species at the post were quite significant. "The valley of the Gila offers favorable conditions for a large number of birds, and forms a very interesting field for the ornithologist." Of special note, he observed "in a mesquite tree . . . a small owl which I have little doubt were of the red-tailed species [Ferruginous Pygmy-Owl]."

In 1907, William L. Sclater, son of the world-renowned English ornithologist Philip L. Sclater, purchased Aiken's bird collection for the Colorado College Museum, which eventually transferred it to the University of Colorado in 1964.

A GALLANT CONFRONTATION, 1876

𝒮𝒪 WHILE ON SICK leave in southern New Mexico Territory in 1876, Lt. Charles Adam Hoke McCauley (1843–1913), a graduate of the U.S. Military Academy with extensive service in the Southwest and Mexico as an artillery officer, made an unusual request. His desire was to join an expedition about to embark for the Red River region of Texas. Although his application was officially denied, the unit commander, Lt. Ernest H. Ruffner of the Corps of Engineers, allowed him to accompany the march as a volunteer.

After leaving Fort Elliott, Texas, near the headwaters of Sweetwater Creek, McCauley (1877) marched southwest into the region of the Staked Plain, where their operations extended for six weeks during May and June. The surface of the plain, approximately four thousand feet in elevation, "is one unvarying level, 'flat beyond comparison', without an object to rest the eye." Short grama grass, yucca, and cactus occur, while in the deeply eroded stream canyons, which ultimately form the Red River, luxuriant vegetation provides "truly oases" compared to the surrounding dry landscape. These drainage systems, known as the "breaks" by the scouts, "are recognizable at great distances by the mirages that can be seen hanging above them."

Crossing the plain proper revealed few bird species, but one McCauley observed almost daily was an abundant aerial "songster," the Horned Lark. While on this march, he gave an unusual but compassionate account of how a nesting lark, in a gallant confrontation, was able to hold the field in front of an understanding army. "One day, during a halt, the column happened to stop within a few yards of a bird upon her eggs, who, after flying to and fro in great solicitude, soon boldly approached, and resumed her place upon her nest with full confidence. The escort was directed to change its course to prevent riding over her, she meanwhile remaining as quiet as if she knew we were friends."

On their return trip, McCauley noted bison were plentiful, as were the "considerable numbers" of Black Vultures attracted to the "many carcasses of animals slain by hunters, generally for their hides alone, very little of the meat being used." Beneath "a huge buffalo hip-bone," he found a nest of a Lark Sparrow, which contained an egg of a Brown-headed Cowbird.

Along the streams McCauley encountered handsome Scissor-tailed Flycatchers with long tails trailing freely in flight; acting as "two elegant feathery tines [that] cross and open at volitiou, whence the ordinary similarity to a pair of scissors." Addressing their habits, McCauley added:

> These birds are grace itself when on wing, darting here and there
> as quick as thought, in buoyant sweeps and curves. The delicate crim-
> son below their wings, as they go glancing by, glows in contrast with
> the beautiful hoary ash of their general plumage; and as the little
> heart ceases to palpitate, you pick up your specimen with a pang of
> remorse, and for once mentally agree with the friend beside you—
> visiting the Staked Plain in the "invalid" interest, and strongly anti-
> collector—that, as he avers, "a bird-skinner is as a butcher." Even
> the teamsters call them "mighty pretty", and no one wonders that the
> "Texicans" . . . brag on their beauty, and call them "Birds of Paradise".

The notes of McCauley record 102 bird species. They are written with polished refinements reminiscent of many similar embellishments scribed in the works of his editor, Elliott Coues, one of the great masters of ornithological verbiage.

THE RULING PASSION OF HIS LIFE, 1876–1878

FORT BROWN ON the Rio Grande River at the southern extreme of Texas has been an important ornithological location for many naturalists. Assistant surgeon James Cushing Merrill (1853–1902) was one of those individuals whose interest in birds, while serving at remote posts, "deepened and broadened until it became, next to his profession, the ruling passion of his life" (Brewster 1910). Beginning in 1876, he served thirty months of his twenty-seven-year army career at this outpost. Although he married later, he was accompanied to this post by his devoted mother, who shared not only his interest in science, but the isolation and limitations that attend this life style.

Merrill (1876, 1878) recognized that the region offered an "excellent field for the ornithologist" because of the river valley, the number of migrants, and its close proximity to the tropics. He found the region rich indeed, for he added several new birds to the avifauna of the United States.

Among them was the long-tailed Common Pauraque, which through its nocturnal "characteristic notes" caught his attention. Others include the Northern Jacana, the small Least Grebe, the tiny red-eyed Yellow-green Vireo, and in all probability, he was the first to observe the "beautiful" Varied Bunting.

Another was the Bronzed Cowbird which, because of its "habit of fluttering in the air," first attracted his attention. The red-eyed male cowbird, with an all black shiny plumage, begins its courtship by fluffing its feathers about the ruff of its nape and then, throwing back its head, it struts stiffly with quivering wings and drooping tail. Merrill (1877b) observed these birds "eagerly paying their addresses. . . . Every now and then one of the males rises in the air, and, poising himself two or three feet above the female, flutters for a minute or two, following her if she moves away, and then descends to resume his puffing and bowing." This unique behavior may cease, repeat, or end in excited copulation with the female. Merrill discovered eight birds that were parasitized by this species. Expanding westward, the Bronzed Cowbird extended its range to the lower Colorado River by 1950 (Monson 1954).

The tropical trochilids from the Americas, considered by the European collectors as novelties, were quickly described as they became available. Merrill (1876, 1877a) found two species rare north of the border, the Rufous-tailed and the Buff-bellied Hummingbirds. Of the 427 species of trochilids listed in the collection of the National Museum to this date, thirteen were found to occur north of the border (Ridgway 1880). Later taxonomic refinements would reduce the total number of hummingbird species by almost a hundred and few were yet to be added to our fauna.

William Brewster (1851–1919), president of the Nuttall Ornithological Club for over forty years, maintained a lasting friendship with Merrill that began when they were school boys and continued through their adult years. Although Brewster (1910) had many kind thoughts about Merrill, including his professionalism and contributions toward ornithology, one especially noble characteristic deserves mention. "He was so wholly superior to jealousy, prejudice and worldliness, and to all considerations of selfish policy, that he never seemed to suspect their possible existence in others." Merrill gave his cherished bird specimens to Brewster, who ultimately left them, along with his own fine collection, to the Museum of Comparative Zoology at Harvard College (Henshaw 1920b).

It was through a long friendship with Charles E. Bendire that Merrill's collection of nests and eggs were ultimately placed in the National Museum. He was also asked to join in the organization of the American Ornithologists' Union but, because of military obligations, he was unable to attend or become one of its founders.

THE PENNSYLVANIA INDUSTRIALIST, 1877–1887

George Burritt Sennett (1840–1900) was certainly an energetic man endowed with many talents and abilities. A Pennsylvania businessman and manufacturer of oil-well machinery by profession, he also became keenly interested in birds at the age of thirty-four. This special interest developed after the purchase of *Field Ornithology* written by Elliott Coues (1874b). Sennett's enthusiasm grew following his subsequent correspondence and later acquaintance with Coues.

Sennett's (1878) entry into the Southwest began in 1877 after thoughtful consideration of just where he might make a serious contribution to ornithology. "Last winter, having inclination and leisure to prosecute the study of birds in a more extended field . . . I began to look about for a suitable location. As is always the case when real desire for study arises, avenues of investigation opened in all directions; but the weight of influence drew me to the Rio Grande."

Three trips were made by Sennett, all in the spring, with F. S. Webster as his assistant in 1877, J. H. Sanford in 1878, and J. M. Priour in 1882. W. Lloyd (1887) worked independently and also collected for Sennett in west Texas in 1887. Several papers were published by Sennett, but the full extent of his anticipated work was never completed because of his untimely death. The notes of his first trips were included in the *Bulletins* of the Hayden Geological and Geographical Surveys of the Territories, edited by Coues.

In addition to the southern coastal region of Texas, Sennett's main objective was to explore from the mouth of the Rio Grande to a few miles above Hidalgo, a distance of approximately three hundred river miles. Starting on March 20 and lasting two months, Sennett's (1878) first trip began with conveyance problems coupled with considerable "annoyances." The latter consisted of "intensely hot days" in which the birds would sometimes spoil before their preparation and, among the less desirable creatures, the unexpected appearance of "huge rattlesnakes." He continued despite considerable discomfort inflicted by his insect antagonists, which would have caused most to falter. "[M]ore troublesome enemies were . . . wood-ticks and red bugs, to say nothing of the fleas. The wood-ticks we could pick off or dig out, but the abominable 'red bugs' . . . too small to be seen, worked themselves through the clothes and into the skin, making one almost wild with intense itching. We only obtained partial relief by giving ourselves, from head to foot, before going to bed, a bath of ammonia, and a daily bath of kerosene oil before going into the brush."

Although Sennett encountered many difficulties, he enjoyed his stay in Hidalgo and had moments of avian delights. During his routine evening bath in the Rio Grande, he never failed to hear the "sweet melody" of Yellow-breasted Chats, considered by him to be "by far the finest singers of all our birds." His great pleasure in hearing their nocturnal song gave him good reason to remark affectionately that "No matter at what time we might wake on a still night we could hear 'our chats.'"

Sennett's first year's work yielded 150 species, of which four new species were added to our fauna north of the border. The first three included the Tropical Parula, Couch's Kingbird, and Brown-crested Flycatcher. The flycatcher was first collected by Merrill near Fort Brown and was not reported until Sennett noted it at Hidalgo.

The White-tipped Dove *(Leptotila verreauxi)* was another bird entering the southern limits of the United States that Sennett (1878, 1880) collected for the first time in this country. In the following year he also collected the eggs. In 1855, Charles Bonaparte named this dove for two French brothers, Jules P. (1807–1873) and Jean B. (1810–1868) Verreaux, who collected and traded natural history specimens from throughout the world. Spencer Baird also acknowledged that J. P. Verreaux supplied many specimens from Mexico and Guatemala that were used for "comparison" with "closely allied species" found in the United States (Baird, Cassin, and Lawrence 1858).

The following April and May, Sennett worked the same region except for a three-week period in which he was incapacitated because of a "poisonous thorn" in his knee. This too was a productive season with four new birds added north of the border from a collection of 167 species. Many birds were repeats, but he was able to gather additional data on their life histories.

Among the flycatchers Sennett (1880) procured was the Great Kiskadee, a large insectivore that he found to be "excessively garrulous" in song during the breeding season. He also observed them "dive into the water [without submerging] after water-insects and minnows that were swimming near the surface." The diminutive Northern Beardless-Tyrannulet, considered by Sennett to be the "best find" of the trip, was obtained at Lomita. He also added another species, the Groove-billed Ani, which was discovered in Mexico by William Bullock.

Sennett devoted over two pages to his experiences collecting White-faced Ibises in an immense salt marsh near Brownsville. In describing the wild cacophonous scene, he wrote: "I was so completely overwhelmed by the sight above and about us, that I was for the time transfixed. A hundred acres of beautiful birds, plunging and screaming above the rushes! Just think of it! . . . On every side were nests in great numbers, and birds guarding their eggs or young."

In 1877, Sennett, in company with Merrill, first mistook a pair of White-tailed Hawks for Ferruginous Hawks after discovering their nest near Brownsville (1878). The following year, however, he readily admitted his mistake after obtaining a specimen at Corpus Christi in March. In May of the same year, Merrill wrote Sennett that he also collected a set of eggs and a specimen near Brownsville (Sennett 1880). In 1882, Capt. Benjamin F. Goss (1823–1893), who Sennett also met the same year, found the species "breeding abundantly" in "a strip of open bushy land lying between the thick line of timber and chaparral along the coast and the open prairie" near Corpus Christi (Bendire 1892).

Joel A. Allen (1901), a founder and first president of the American Ornithologists' Union, wrote a tribute to his friend and acquaintance: "Mr. Sennett . . . impelled by the instincts of a true naturalist, has left his mark upon the progress of American ornithology, and has contributed not a little in the way of 'bricks and straw' to the construction of that edifice, for the perfection of which we are all lending our efforts, each in proportion to his opportunities and endowments."

SOUTHERN TEXAS FASCINATION, 1882–1900S

A HOST OF early ornithologists following Merrill and Sennett have contributed in a significant way to the study of Texas birds. Among the several collectors were Nathan C. Brown (1882, 1884) in Kendall County, J. Douglas Ogilby (1882) in Navarro County, and Joseph L. Hancock (1887) in the vicinity of Corpus Christi. Charles W. Beckham (1856–1888), in an effort to gain a cure for an illness from which he eventually died, traveled west and spent four months of his young life accumulating data on the birds of southwestern Texas. Beckham's (1887) work, also noting significant indicator plants of the areas he explored, provided an annotated list of 283 species, including those of other observers who collected in the same localities.

Frank M. Chapman (1864–1945) undertook a short spring visit to the Corpus Christi area and subsequently published a paper in which he discussed his work in relation to Beckham, giving a comparative analysis of their observations with respect to each other (Chapman 1891). As an ornithologist, Chapman served with the American Museum of Natural History for over half a century, exerting his influence with regard not only to the classification and habits of birds, but also to their protection. He referred to birds as "Nature's most eloquent expression of beauty, joy and freedom" (Murphy 1950). Like his predecessor, J. A. Allen, Chapman's (1894) ornithological interests were wide and included such subjects as the influences that prompt bird migration.

The configuration of North America is conducive for most birds on their fall migration to pass through Mexico to Central America and, for some, to continue even to South America. In spring the reverse is generally repeated, thus avoiding the precarious water expanse of the Gulf of Mexico. Still, others fly non-stop over the Gulf, confronting unpredictable and severe storms. Abbott Marston Frazar (1860–1925), a naturalist returning east by ship along the Texas coast, witnessed a tragic event on this enormous obstacle; an avian spring calamity at sea (1881):

> *April 2, found me in a small schooner, on the passage from Brazos de Santiago, Texas, to Mobile, Alabama. At about noon . . . the wind suddenly changed from east to north, and within an hour it was blowing a gale; we were now about thirty miles south of the mouths of the Mississippi River . . . on a line with . . . the peninsular of Yucatan. Up to the time the storm commenced the only land birds seen were three Yellow-rumped Warblers . . . but within an hour after the storm broke they began to appear, and in a very short time birds of various species were to be seen in all directions, singly and in small flocks. . . . These birds of course must have been far overhead and only came down near the surface of the water in endeavoring to escape from the force of the wind. By four it had come to be a serious matter with them, as the gale was too strong. . . . As long as they were in the trough of the sea*

the wind had very little effect . . . but as soon as they reached the crest
of a wave it would catch them up and in an instant they were blown
hundreds of yards back or else into the water and drowned. . . . It was
sad indeed to see them struggling along side of the vessel.

Frazer recognized at least twenty-three species in this incident.

In addition to working for William Brewster on the Mexican mainland, Frazar also collected for him in Baja California during 1887 (Palmer 1926). His forty-four hundred specimens taken in that region form the basis for 255 species discussed in Brewster's "Birds of the Cape Region of Lower California" (1902).

Henry Philemon Attwater (1854–1931), an English emigrant, compiled a list of 242 species in the vicinity of San Antonio over portions of five years (1892a). During an early spring migration, Attwater (1892b) reported a three day "norther" in which thousands of warblers consisting of six species perished at Rockport on the Gulf of Mexico.

Beginning in 1888, Frank Blake Armstrong (?–1915), a resident of Brownsville, collected, often with his wife, thousands of specimens in the lower Rio Grande region (Griscom and Crosby 1925, 1926). They were taken mainly on the Texas side of the border and were sent to many museums throughout the country, including the Philadelphia Academy of Natural Sciences. In 1892, Armstrong collected a Mexican species new to the United States, the Golden-crowned Warbler.

This border country also attracted another easterner, Samuel N. Rhoads (1862–1952) from Philadelphia, who briefly visited Corpus Christi during the breeding season in 1891 en route to Tucson. With his list containing 108 species, he also noted from the descriptive data of previous investigators the "comparatively sudden conversion of many square miles of debatable ornithological ground from prairie to brush-land" (Rhoads 1892).

The culminating work on ornithology in Texas was the two-volume work *The Bird Life of Texas* (1974) by Harry Church Oberholser (1870–1963), edited by Edgar B. Kincaid, Jr. Begun in 1900, this exceedingly detailed work consumed almost his whole life. As an ornithologist recognized for his "marked propensity" toward "taxonomic splitting," Oberholser contributed profoundly to the literature with over nine hundred publications (Aldrich 1968). While serving almost half a century with the Biological Survey and its successor agencies, his work also extended to the borderland states of New Mexico and Arizona.

A FORTUITOUS NATURALIST, 1875–1900S

🕉 WITH A PAIR of mules and a spring wagon as a means of conveyance, Frank Stephens (1849–1937), accompanied by his bride of a few months, journeyed west to Colorado Springs in 1873. With a strong interest in natural history, but a limited knowledge about "stuffing" birds, Stephens became acquainted with Charles Aiken, who, in agreeing to purchase some specimens, taught him the skills of making bird skins.

Stephens (1879) then moved south to the mining district around Silver City in southwestern New Mexico Territory in 1875 and 1876. Among the several birds he noted in the area was a nesting pair of Zone-tailed Hawks in a cottonwood grove on the upper Gila River.

In 1876, when the Chiricahua Apaches living in southeastern Arizona Territory had their four-year-old reservation terminated, they were removed to the San Carlos Reservation on the Gila River under the watchful eye of the army. To avoid forced relocation, many Apache scattered to various parts of southeastern Arizona and southwestern New Mexico Territories, while others, including Geronimo (ca. 1823–1909), fled south into adjacent Mexico.

Over thirty years later, Stephens, in reminiscing about the dangers of that year with his sixteen-year-old southern California protege Laurence M. Huey (1892–?), told of how he used a "gun loaded for men, not for birds" to collect a Northern Pygmy-Owl. "See here, I shot that bird with a [.]50 caliber Henry rifle. It was the first Pigmy Owl I had ever seen, and a very dangerous chance I took to get it, to, for the Apaches had raided a ranch next to mine, only five miles away, and we were hurrying to the shelter of Fort Bayard. In all probability the Indians heard the shot, but they never caught up with us, nor did they attack the fort" (Huey 1938).

Stephens' (1918) decision to continue to southern California was prompted by another precarious situation. "In the summer of 1876, the Apaches were troublesome, with prospects of worse times ahead, and as we were living on an exposed mountain ranch, we decided that we had better get out. The Indians had stolen the horses I had traded my mules for, but I bought a yoke of oxen and started on for California. We were fortunately not molested on the way, but settlers were killed ahead of us and after we passed."

They stopped for a month at Tres Alamos, a stage station forty-five miles east of Tucson near the San Pedro River, before passing through Tucson, the Maricopa Indian villages, and on to the Colorado River. Yuma was a "comparatively busy place" with mails passing every four days via six-horse stages. Freight arrived by river steamers for distribution to army garrisons and mining camps.

In 1880, Stephens was lured back from California for a short period to the mining camps around Tombstone. Then, in March of the following year, he collected for William Brewster on a five-month trip that began on the rugged eastern slopes of the Chiricahua Mountains and ended in southern California. Galeyville, his starting point, had just experienced a mining boom, and he was apparently among those departing almost as quickly as they arrived. Apache and troop movements continued between San Carlos Indian Reservation and Mexico until Geronimo's final surrender in 1886. Also, in 1881, the Southern Pacific Railroad completed its track from Texas across southern New Mexico and Arizona into southern California.

Immediately prior to leaving the Chiricahua Mountains for California, Stephens shot two Wild Turkeys, a species not recorded among the 165 received by Brewster. Huey (1938) wrote of a quotation given him by Stephens, "I was just leaving my cabin, which was situated about a mile up the canyon west of Galeyville . . . to cut a load of wood. . . . In that country, in those days, no one ever left his cabin without his rifle. I was only a few rods from the cabin when two large turkeys ran across an opening in the trees just ahead of me. . . . I got both of them. Mrs. Stephens, hearing the two reports, thought someone had shot me and came running out of the cabin. The joke was on her though," Stephens added in good humor, "and as a penalty she carried both of the birds back to the cabin. They were big fellows, too!" The Wild Turkey was extirpated from the Chiricahua Mountains around 1909.

Stephens briefly explored Cave Creek on the eastern flank, then skirted northward around the Chiricahua Mountains to Camp Bowie before proceeding southward to Morse's Mill near Turkey Creek on the western slope. Most of the canyon streams are ephemeral, but while at the lumber mill, Stephens noted a bird rare to these mountains by a brook. "My attention was called to the song of some bird which came from the mountain brook running past camp. There was a steep, rocky wall on the further side, and the notes echoing from it, and mingling with the purling of the water, sounded exquisitely sweet. On looking for the author, I noticed some ripples rolling out from behind the willows that fringed the nearer shore, and soon discovered an Ouzel [American Dipper] dabbling in the shallow water" (Brewster 1882).

Stephens was the second naturalist, after Henry Henshaw, to enter the higher conifer forests of the Madrean Archipelago north of the border and the first to venture into the Chiricahua Mountains. His entry into these mountains resulted in the introduction to our fauna of the Mexican Chickadee *(Poecile sclateri)* (Brewster 1881). In 1857, English ornithologist Philip L. Sclater (1829–1913), in describing the onomatopoeic Mexican Chickadee taken by Auguste Sallé in southern Mexico, unknowingly applied a name for this species which was preoccupied by another bird. German ornithologist Otto Kleinschmidt (1870–1954) when learning this, honored its original describer in 1897.

The Chiricahua and nearby Animas Mountains are the only ranges north of the border where these charming birds occur. Interestingly, the high Huachuca and Santa Rita Mountains, immediately to the west, are not represented by the presence of any chickadee (Marshall 1957; Phillips, Marshall, and Monson 1964). However, the congeneric Mountain Chickadee occurs immediately north in the Graham, Rincon, and Santa Catalina Mountains, which includes its southern range limit extending from west Texas to California and into the northern Baja Peninsula. Nowhere do the two species overlap during the breeding season.

Continuing west from Morse's Mill, Stephens crossed the Sulphur Springs Valley to the bustling mining town of Tombstone, then, turning northward, he stopped briefly at Cienega Station before traveling down Pantano Wash to Tucson. This area "proved so rich in desirable birds" that he prolonged his stay for two months. During this period Stephens also made a trip south into the Santa Rita Mountains, where he added the Dusky-capped Flycatcher, another Mexican species, to our fauna north of the border.

The July crossing of the "arid plains and scorching deserts" of southwestern Arizona and southeastern California by Stephens "was attended with such privations, and often positive suffering, that little attention could be paid to birds." Relief was finally achieved when he arrived in Riverside, California, his final destination.

Before returning to Arizona, Stephens (1902) became the first naturalist to climb to the high forested San Jacinto Mountains (10,786 feet) of southern California. While in lush Round Valley (9,200 feet), he collected a tiny Northern Saw-whet Owl.

In March 1884, Stephens again returned to Camp Lowell adding the Blue-throated "Cazique" or Hummingbird to the United States fauna (Brewster 1885). This hummingbird was among the specimens that Paolo Botta gave to the French nobleman Francois Victor Masséna after returning from the Pacific Coast in 1827 or 1828. Messéna sent it along with other specimens to his friend

René Primevère Lesson, who, after having already designated Anna's Hummingbird for Masséna's wife, named this Mexican species for his own artistic second wife Marie Clémence (ca. 1796–1834) in 1829.

In 1893, Edward W. Nelson, while working in south central Mexico, secured the first nest of the Blue-throated Hummingbird. George F. Breninger (?–1905), however, working for the Chicago Field Museum of Natural History, found the first nesting site north of the border in "the box" of Ramsey Canyon in the Huachuca Mountains. It was "fastened to the longest fronds" among the "prettiest masses of maiden-hair fern that I had seen since leaving the Pacific Coast" (Breninger 1899). An excellent field naturalist, Breninger apparently died from the effects of arsenic accumulation acquired while preparing specimens over many years.

In August 1884, Stephens (1885) journeyed from Tucson to the Gulf of California with C. G. Pringle, a botanist who was taking his second trip of the year. Before reaching the Mexican custom house at Sasabe, Stephens made several unsuccessful attempts at securing an "unknown Partridge said to occur in this region." Only partial glimpses made its identification uncertain but, he did record the "plainly heard" notes of "bob-white; the bob was as loud as the white." Crossing over to the Mexican side, he became the second naturalist to secure a "Masked Bobwhite," a darker form of the Northern Bobwhite.

Their journey to Puerto Lobos, a place of no "habitation nor inhabitant," was along a desolate road that had not been used since Pringle's former trip five months earlier. On viewing from the shore a scene of remarkable differential tides, which included abundant marine animals and birds, Stephens exclaimed, "What a grand field, although a very difficult one, this Gulf and its shores present for scientific exploration!" After traveling over five hundred miles they returned to Tucson September 1.

In 1902, Stephens (1903), working for the U.S. Biological Survey, recorded 120 bird species during four months on the "hot" desert region common to southeastern California and western Arizona. Leaving San Gorgonio Pass in the middle of May, he traveled a loop that included the oasis of Twenty-nine Palms, the rugged Providence Mountains, the Colorado River near Fort Mojave, the isolated Hualapai Mountains, followed by the Big Sandy River before turning southward to Ehrenberg (119° temperatures) on the Colorado River, and home by way of Chuckawalla Spring. Three years later, Ned Hollister (1876–1924) was engaged by the Biological Survey to explore the Fort Mojave region (Hollister 1908).

While at San Ysabel Rancho in southern California, Stephens (1899) recorded an interesting incident, although not isolated in ornithological literature; the lassoing of a California "Vulture" or Condor. After gorging on carrion, condors apparently became flightless, enabling local vaqueros to throw a "riata" over the sluggish birds.

Although no mention was made of each other, James B. Dixon (1937), in San Diego County where Stephens was residing, engaged in a remarkable thirty-six-year account of Golden Eagle populations. Beginning in 1900, he determined that twenty-seven breeding pairs consistently occupied 945 square miles in an area of coastal mountains.

In 1910, Stephens accompanied Joseph Grinnell down the Colorado River and, over a decade later, he traveled three times to the Baja Peninsula with Laurence Huey. He eventually settled close to San Diego, where he contributed substantially to the collections of the San Diego Society of Natural History. By 1917, Stephens and his wife were devoting most of their time to the development of the museum and, seventeen years later, the title of curator emeritus was bestowed on him.

When telling about their shared field experiences, Huey (1938) understood the pleasures of a congenial relationship. "As a camp companion I believe Frank Stephens never had a superior; there was always a cheery word and a willing hand for any task great or small. His tricks of camplore seemed inexhaustible. If there was a simpler means or a short cut to any outdoor work, or a way to do a better job, Frank Stephens knew it. These things he had learned from his pioneering experiences, from a long life of contact with his fellow man, and from matching his wits against the wiles of nature."

THE CAVALRYMAN, 1884–1886

GENERAL GEORGE S. CROOK (1828–1890), commander of the Department of Arizona from 1882 to 1886, realized early on that if the campaign against the renegade Apaches from the San Carlos Reservation was to be a success, friendly Apaches would have to be employed as army scouts to trail the "outlaw" bands deep into the rugged Sierra Madre of Mexico. On several occasions during this conflict, the general's aide-de-camp, Captain John G. Bourke (1846–1896), recorded Apache superstitions, including their fear of owls, the most "dreaded" of all southwestern denizens. On one occasion, Bourke wrote that the Apache scouts "made a great fuss and would not be pacified until one of the whites [Frank Randall, New York Herald journalist] of our command had released a little owl which he had captured" (Bourke [1891] 1971).

Serving in the campaign, Lt. Harry Coupland Benson (1857–1924), a graduate of the U.S. Military Academy and a cavalry officer, added a great deal to our knowledge about the local avifauna. From 1884 to 1886, when time permitted, Benson collected around Fort Huachuca, sending his specimens to the National Museum. From the fort, situated on the lower northern slopes of the Huachuca Mountains, Benson could climb up sycamore-lined Post Canyon through the oaks into the upper pines, where he could gain a magnificent view of sprawling grasslands to the north and east. Crossing the Mexican border in the distant southeast, the San Pedro River flowed northward through the grassy expanse.

Although the brightly attired Elegant Trogon had been noted north of the border by Dr. J. C. Merrill (1878) in the lower Rio Grande Valley and by W.E.D. Scott (1886a) in the Santa Catalina Mountains, Benson took the first specimen in the Huachuca Mountains (Ridgway 1887a).

Two large dominant plants occurring on the grassy plains below the fort were low scrubby Mesquite and Soaptree Yucca, which sprout long flowering stalks in summer. It was primarily in the Mesquite that Benson located over fifty Chihuahuan Raven nests and, on five of their deserted stick platforms, he found all five occupied by Aplomado Falcons (Bendire 1887). These were the last nesting records for this rare falcon in Arizona, while the most recent credible sight records in this area were over fifty years later (Monson 1942; Monson and Phillips 1981). Benson also reported forty-one nests of the Swainson's Hawk, making it the "commonest" raptor in the vicinity.

While on a scouting expedition in pursuit of Geronimo under the command of Captain Henry W. Lawton (1843–1899), Benson collected a great Imperial Woodpecker, the largest picid in the world, about fifty miles south of the border in the pine forests of the Sierra Madre (Ridgway 1887b). In 1913, Albert K. Fisher, who had visited the Chiricahua Mountains a score of years earlier, also noted in a letter to J. Eugene Law that Benson had secured this woodpecker "about eighty miles south" of these mountains in Mexico (Law 1913; Hubbard 1969).

The last men at the turn of the century to record their observations about this now-extinct (Russell and Monson 1998) crested woodpecker were E. W. Nelson (1898), Norwegian C. Lumholtz (1902), and James H. Gaut (1904). It was the inimical actions of man's predisposition to needlessly destroy wild

Imperial Woodpecker,
Campephilus imperialis
(Gould), 1832; now extinct.
(Cassin 1856)

creatures—for food, body parts, plumage, medicinal powers, or simply self indulgence—in combination with habitat destruction that brought about the extermination of this grand species (Smith 1908; Tanner 1964).

Before the creation of the National Park Service in 1916, Congress created several national parks that were first staffed by army personnel. After 1891, much of Benson's career was spent in various capacities of administration in Sequoia, Yosemite, and Yellowstone.

Gen. Benjamin Alvord, a fellow officer and lifelong friend, wrote of Benson's acquaintance: "A typical officer of the 'old' army, Benson was imbued with its spirit and ideals. High-minded, courageous, honest . . . he was frank and outspoken, while trying always to be just and considerate . . . alert of mind, inter-

James Hamilton Gaut (1879–1914) on the U.S. Biological Survey in New Mexico, 1904. (Courtesy of the Smithsonian Institution Archives, Record Unit 7172, Box 9)

ested in people and things, and a student of both nature and books, he was an interesting and delightful companion" (Farquhar 1925).

FROM DESERTS TO ALPINE SUMMITS, 1884–1894

🦚 ꓮFTER PARTICIPATING IN the founding of the American Ornithologists' Union in 1883, Edgar Alexander Mearns (1856–1916) entered the army as an assistant surgeon, and the following year, at his request, was posted at Fort Verde in central Arizona. Excerpts from Mearns's (1886) notes reveal his thoughts as he arrived in March at this remote post that was to be "home" for the next four years. "I first beheld the wide valley of the Rio Verde, with its tortuous stream winding in zigzags, bounded by a fringe of cottonwoods which, at that season, were destitute of foliage or flower. We gazed with keen interest upon the panorama before us, as the driver of our ambulance pointed out in the distance a series of low, whitewashed sheds surrounding a quadrangle and flanked by some adobe walls and haystacks, which was to be our station."

For the next several months, Mearns rode through the nearby canyons and mountains, encountering, along with seasonal changes, many new plants and animals. He wrote of "Nature" bestowing on this scene treasures with "lavish prodigality," though "veiling them from the vulgar gaze" of many and only revealing their qualities to those with interest to discover. After becoming "reconciled" with his surroundings, he remarked that "It was a dismal and desolate outlook truly, but possessed of the beauty of wild loneliness."

As a surgeon and naturalist, he wrote with some reassurance that Gen. George Crook "is particularly interested in my pursuits, and has chosen me to accompany him on two long expeditions through the wildest and least known portion of Arizona" (Hume 1978). In November 1884, one of those trips was realized when he accompanied the general and his staff to the Havasupai Indian village of Supai on the floor of the Grand Canyon. Mearns was among the first naturalists to enter this segment of the Grand Canyon to observe its avifauna (Brown, Carothers, Johnson 1987). Elliott Coues, on his return trip to Arizona Territory, preceded Mearns by three years when he visited the same village in 1881.

The expedition to the Grand Canyon was going well until their descent into the steep-walled gorge. Suddenly, a mule carrying the cooking gear stumbled to the edge of the narrow trail and plunged over the brink into the depths below. From the canyon floor, the advance party, on hearing "a yell of dismay and horror from the packers," looked up as the helpless animal "rolled and tumbled . . . striking everything in its path . . . it struck with the deafening report of a cannon upon a towering mass of sandstone" (Casanova 1968). The most numerous birds he observed were Gambel's Quail, which are seldom if ever seen in this area today.

Prior to arriving at Fort Verde from a long mid-May excursion in 1885, Mearns (1886), with his horse Daisy, shared a "peculiar sensation of pleasure and relief" upon safely reaching a shady cottonwood grove along the New River. Here, in an isolated but refreshing setting, he unknowingly set the scene for his chance encounter with a large, black Neotropical raptor.

> [I] filled my canteen in the stream and drank, while my brute companion slaked her thirst after the manner of her equine kind. We had travelled nearly a thousand miles, and were now within a few days' march of home. Few trees had rested our eyes from the glare of the tropical sun, or had shielded us from the fervid heat of its piercing

rays, upon the scorched desert wilderness that we had traversed. Here was shade, and the sweet sound of a running stream, upon whose margin a handsome nosegay might have been easily plucked. . . . Soon "Daisy" was munching sweet herbage upon the shore, and perchance was thinking of the good barley soon to be enjoyed in her snug stall in the Quartermaster's corral at Fort Verde, whilst my own thoughts had wandered to very nearly the same locality, when both were interrupted by the shrill whistle of a [Zone-tailed] Hawk that came gliding towards me through the dark shadows of the dense foliage.

The following day, Mearns "encamped" on the Agua Fria River, which he portrayed as one of the "prettiest places" he had seen in Arizona Territory.

The rocky sides of the cañon were covered with cacti of diversified shapes, from the gigantic Cereus *[Saguaro] to the* Echinocacti *[Barrel] and* Opuntiae *[Prickly Pear]. Beautiful flowers grow beneath the tall cottonwoods, which here form the handsomest groves that I have yet seen. The cañon echoed the voices of hundreds of feathered songsters, and the hum of insects and countless Hummingbirds filled the air. Flocks of beautiful White-winged Doves drank upon the sandy brink and then betook themselves to the dense foliage overhead, where their loud and mournful cooing filled the air. An occasional glimpse of the gorgeous plumage of the [Northern] Cardinal was obtained, and the shining Phainopeplas darted after insects from the sides of the cañon.*

Mearns's (1890) travels in the mountainous region of central Arizona Territory were extensive. In the summer of 1886, he "encamped on a southern slope of the Mogollon Mountains," where he followed "a small stream into a little cañon between whose rocky walls stood groups of towering spruces and of aspens, the ground beneath thickly sprinkled with violets, strawberries, honeysuckles, and columbines." There he flushed from its nest "the first Red-faced Warbler I had ever seen. . . . I descried a small opening close beside it among the stones and pine needles; on parting some blooming honeysuckles . . . and moss, I discovered the nest,—most artfully concealed."

In 1887, Mearns, with the use of a "hollow log" as a boat, attempted to determine the various waterfowl breeding on several small lakes, including Stoneman's and Mormon Lakes. He found Eared Grebes breeding abundantly on both lakes, but on Mormon Lake, where there was a greater abundance of ducks, the breeding success was quite poor, because the Mormon settlers had been "gathering" eggs "about a week before our arrival." However, he did observe Northern Shovelers, Gadwalls, Northern Pintails, Mallards, American Widgeon, and Ruddy Ducks. He recorded that only Cinnamon Teal were breeding successfully on the lakes, while along some of the mountain streams, he also noted Common Mergansers nesting.

In the summer of 1887, two years before C. Hart Merriam's (1890) eventful survey of the San Francisco Peaks, Mearns climbed to the high alpine summits of those volcanic mountains. "Not far from the timberline," he collected a Northern Goshawk. On that memorable trip, Mearns (1890) became the first naturalist to attain these lofty peaks and experienced his "windiest day" ever.

I succeeded with a companion in reaching the summit [Wheeler Survey, 12,562 feet; current USGS, 12,633 feet] of Humphrey's Peak— the highest land in Arizona—and crawled up behind the monument built by Lieutenant Wheeler's party. We looked down upon the remaining half-dozen [cinder] cones . . . which constitute the San Francisco group . . . snow-like masses of clouds rushed past us, and the pile of rocks behind which we were cowering vibrated in a gale so fierce that large pieces of volcanic scoria, thrown into the air, were swept along over the brink of a precipice in front of us. The only living things in sight, besides some [Bighorn] mountain sheep, were two birds, in point of size inclining to extremes of the ornithological scale,—a Prairie Falcon, and a Broad-tailed Hummingbird, which later sought momentary shelter with ourselves. These only, braved the wind and cold at the summit.

While in northern Arizona Territory, Mearns developed a phonetic ornithological vocabulary (1896) spoken by the Moki [Hopi] Indians. In this work, he noted Captain Bourke's familiarity with the Indians in regard to their use of "two species of [Golden and Bald] eagles, which [were] kept by them in cages" for feather trade and ceremonial dress.

From 1892 to 1894, Mearns was attached to the United States–Mexico International Boundary Commission, which was under the direction of Col. J. W. Barlow. The commission was to repair and reestablish the survey monuments along the border from El Paso to the Pacific Ocean. In his position as medical officer, he, along with his assistant Frank Xavier Holzner, was able to establish "without additional cost" a biological section to collect the flora and fauna (International Boundary Commission 1898).

The principal work was carried out during the devastating drought of 1890–1893. Hazards were still considerable. In the San Luis Mountains (6,719 feet) of extreme southwestern New Mexico Territory and adjacent Mexico, special care was taken because of "numerous Sonoran grizzly bears and . . . the 'Kid' and his band of renegade Apache Indians added danger" (Mearns 1907). Nearby at San Bernardino Springs in Arizona Territory, Mearns "was prostrated by aestivo-autumnal malaria fever."

One occasion around midnight, twenty miles east in Cajon Bonito Creek on the Mexican side of the border, "rushing torrents" of water from a distant cloud-burst in the San Luis Mountains brought panic to their sleeping camp. Except for the packers and animals encamped on the bluff, "the rest of the party only saved themselves and their belongings by hastily suspending their guns, bedding, saddles, etc., to the limbs of a huge sycamore tree into which they climbed. . . . The volume of water was surprisingly great as no rain had fallen near us . . . only the sycamore stood above the flood, appearing to our packers . . . much 'like a Christmas tree.'"

Arriving in the Coast Range of southern California, Mearns took special note of the winter food of the local Indians. "Probably no human beings are so dependent on the kind offices of birds . . . for when the mountains are deeply covered with snow . . . the Indians are obliged to subsist almost wholly upon the acorns which the more provident [Acorn] woodpeckers have stored in the bark of the pine trees."

As a closing to his fieldwork in the Southwest, Mearns joined A. W. Anthony, another prominent naturalist, on a trip to San Clemente Island off the coast of southern California. Over thirty thousand plant and animal specimens, including eight thousand birds, were secured on this boundary survey and sent to the National Museum.

THE LAST OF ITS KIND, 1885–1891

Dr. Robert Wilson Shufeldt (1850–1934) was posted at Fort Wingate, New Mexico from 1885 to 1890. While there, he found sufficient idle time and interest to accomplish many bird studies in the areas of osteology, morphology, myology, and later in the east, paleontology (Lambrecht 1935). A man of untiring energies, he also sought out knowledge on the natural history of birds.

Shufeldt (1887b) frequented a "magnificent gorge" just three miles west of the post, from which rose 350-foot perpendicular walls. "It was within the deep and crack-like fissures seen in the walls of the eaves of these latter recesses, away high up on either side of this rocky chasm, that [White-throated Swift] resorted to lay its eggs. So wisely had every pair of these birds chosen the cleft wherein their nests were hidden, that all my plans and attempts to secure a set of eggs proved futile." Although not the first oologist to secure the eggs of the White-throated Swift, Wilson C. Hanna (1883–1982) was the first to discover their torpidity during an "extremely cold wave" on Slover Mountain in southern California (Hanna 1909, 1917).

Shufeldt (1887a) was probably the first naturalist in the Southwest to use a camera to photograph wild birds in their natural setting. The idea to him was most fulfilling. "I know of no pursuit so thoroughly full of interest for the ornithologist as this photographing of birds in their native haunts. It requires, too, all the ingenuity at our command, to say nothing of patience, to pursue it successfully." Shufeldt also promoted the use of the camera for photographing bird eggs.

Shufeldt was one of the founders of the American Ornithologists' Union and became an honorary curator at the Smithsonian Institution. In this latter capacity, Shufeldt (1915) wrote an account of his examination of the last Passenger Pigeon. Birds were still observed at a roost in the Southwest near Austin, Texas, as late as 1881, but shortly after the turn of the century "the last, the very last, of millions upon millions" of this wonderful feathered species was extinct.

7 THE DWINDLING FRONTIER

"The remnant of that picturesque and interesting company [Thick-billed Parrots], concluding perhaps, though wrongfully, that they were unwelcome to citizenship in this great republic, disappeared, returning, probably, to the land whence they came; and if they tell hard things of the inhabitants of Arizona to their fellows in that country, and to such of its human inhabitants as speak their language, they can scarcely be blamed."
—RICHARD D. LUSK (1900)

WITH THE EXPANSION westward came the realization that certain native faunal species were not numerically infinite, but were, in some instances, most assuredly in jeopardy, with a distinct possibility of extinction. From this period emerged the first efforts to establish professional scientists within government agencies. Societies consisting of both scientists and laity were also formed in the private sector, to study species and develop strategies for their protection, or in some cases, their elimination.

SCIENCE, EDUCATION, AND PROTECTION, 1883–1900S

AFTER A TEN-YEAR existence, the Nuttall Ornithological Club, primarily a local bird organization in Cambridge, found that its future was tentative and somewhat in doubt. The leadership—William Brewster, president; J. A. Allen, *Bulletin* editor; Elliott Coues, associate editor—in viewing the Club's prospects for survival, recognized that in its stead, there was a need for a similar, but larger national organization.

In 1883, the same three men, using the Nuttall Ornithological Club as a medium, invited a select group of just over fifty ornithologists to convene in New York City to form the American Ornithologists' Union. The invitation listed several items for consideration, but the pivotal thrust of the proposed convention was "the promotion of social and scientific intercourse between American ornithologists, and their co-operation in whatever may tend to the advancement of Ornithology in North America" (Brewster, Allen, and Coues 1883). Not all members of the Nuttall Ornithological Club were enthusiastic about the methods, or the secrecy, that almost brought about the sudden demise of their club; nevertheless, the American Ornithologists' Union (AOU) was launched with high expectations.

S. F. Baird and J. A. Allen, who were unable to attend the meeting, were included with the twenty-three attending men to become the founders of the new organization. Of these, naturalists who had experiences in the Southwest were well represented by the following: H. W. Henshaw, A. K. Fisher, E. A. Mearns, C. Hart Merriam, C. F. Batchelder, C. E. Bendire, E. Coues, and R. W. Shufeldt. In addition, W. Brewster and R. Ridgway were directly associated with the field naturalists in various ornithological endeavors. The first-year officers were Allen, president; Coues and Ridgway, vice presidents; and Merriam, secretary-treasurer. The founders also added twenty-one additional members of which the following six were southwestern naturalists: L. Belding, J. G. Cooper, J. C. Merrill, E. W. Nelson, W.E.D. Scott, and G. B. Sennett. G. B. Grinnell, R. Deane, and G. N. Lawrence were also associated in supportive roles. Of the original forty-seven members, almost one third were naturalists who had actively explored in the borderland region of the Southwest.

In 1884, the first volume of the *Auk* was published. Of the several committees formed since the inception of the AOU, the Committee on Classification and Nomenclature has emerged to address one of the primary concerns of this professional scientific organization.

This was also a period in which the media, namely *Forest and Stream* along with many newspapers, published a number of articles about the slaughter of birds for food and the millinery trade. From this awareness sprang almost instant public support for bird protection.

During the next two years, the AOU addressed these and many other problems related to the science of ornithology. The AOU also formally appealed to Congress for the creation of a government agency to coordinate data gathering on bird migration. In 1885, Congress appropriated monies to accomplish this within the Department of Agriculture, and C. Hart Merriam was appointed to direct the task. Food habits of certain birds were also studied with regard to agriculture economics, and this resulted in efforts to suppress or eliminate certain species. From these beginnings emerged the Bureau of Biological Survey. The agency later assumed many tasks of exploration, protection, and the overseeing of congressionally designated reservations that were specifically established for species conservation. Over the last century, after a series of name and mandate revisions, the organization evolved into the U.S. Fish and Wildlife Service. During the same period, Congress was beginning to address land and resource policies, which eventually led to the creation of the Forest Service, the Bureau of Land Management, and the National Park Service.

In 1886, growing public sentiment against the fashionable practice of wearing bird feathers encouraged the development of the Audubon Society. The original society was named in honor of John J. Audubon by author, ethnologist, and naturalist George Bird Grinnell. As editor of *Forest and Stream*, Grinnell attempted to lead the effort to enlist public support by printing an editorial which read in part: "In the first half of this century there lived a man who did more to teach Americans about birds of their own land than any other who ever lived. His beautiful and spirited paintings and his charming and tender accounts of the habits of his favorites have made him immortal, and have inspired his countrymen with an ardent love for the birds. The land which produced the painter-naturalist, John James Audubon, will not willingly see the beautiful forms he loved so well exterminated" (Dutcher 1904). The "First Audubon Movement" rose quickly in 1886, but it rapidly ebbed and became virtually nonexistent at the end of two years.

The AOU also became frustrated in its attempts to get proper legislation passed and to disseminate information relative to bird protection. Then, in 1895, a new AOU Committee on Protection of Birds, under the chairmanship of William Dutcher (1846–1920), came to the conclusion that making the national movement a success would require a cooperative effort consisting of a system of state Audubon Societies (Palmer 1921). Thus, under the leadership of Dutcher, the "Second Audubon Movement" began with a broader base to educate the public about birds and the need to protect them. In 1899, *Bird-Lore* began its bimonthly circulation as the official journal of the new organization. At the same time, the AOU was redoubling efforts at bird protection by promoting education and the adoption of "State Bird Laws" (Stone 1899a). As they did on the rest of the nation, these organizations and their publications had a profound effect on the Southwest.

By the turn of the century, congressional action had begun to protect portions of the vast natural resources of the nation. In 1893, Grand Canyon Forest Reserve was established by President Benjamin Harrison. Presidents Grover Cleveland, William McKinley, and Theodore Roosevelt followed in quick succession with several more southwestern forest reserves. The Chiricahua Mountains, typical of the border ranges, had eleven saw mills in operation, which, by 1902, had cut a third of its forest (Bahre 1991). As settlement increased, the need to protect other areas became apparent.

In 1906, Roosevelt, a champion of the environment, established through the Antiquities Act the first three national monuments in the Southwest: Petrified Forest, Montezuma Castle, and El Morro. While attending Harvard, Roosevelt had become a member of the Nuttall Ornithological Club. Charles Foster Batchelder (1856–1954), a leading officer in the club and a visitor to the Southwest, recognized the youthful Roosevelt for his abilities, but felt that he "seemed a bit too cocksure and lacking in the self-criticism that, in our eyes, went with a truly scientific spirit" (Taber 1958). This was doubtless a fair appraisal of "Teddy" at the time, however, Roosevelt (1924) later added "while my interest in natural history has added very little to my sum of achievement, it has added immeasurably to my sum of enjoyment in life."

AVIAN PERDITION, 1891–1900S

Two avian species were observed at their western and southern continental limits by several Texas naturalists and others, but they then disappeared at the turn of the century before much of their life histories had been recorded (Wright 1912; McKinley 1964). The gorgeous Carolina Parakeet was already diminishing and nearly extinct when Edwin M. Hasbrouck (1891a) predicted its ultimate fate would be at "the ruthless and wanton destruction wielded by the hand of man." The last wild specimen was taken in Florida in 1914. With a similar distribution, but with survival circumstances generally related to habitat loss, the large, crested Ivory-billed Woodpecker was gradually eliminated from its former range in southern and eastern Texas and ultimately on the continent (Hasbrouck 1891b).

DETERMINED FIELD ORNITHOLOGIST, 1881–1885

🌀 IN 1881, WILLIAM Earl Dodge Scott (1852–1910) obtained a leave of absence as Curator of Ornithology at Princeton College Museum of Biology to visit southeastern Arizona. Scott took extensive notes on the birds of the Santa Catalina and Pinal Mountains, including the surrounding environs of the Santa Cruz and San Pedro Rivers. During the next four years, he listed 245 species and collected over 2,500 specimens, which he sold to the American Museum of Natural History.

Scott was a man of sheer determination in his pursuit of ornithology despite a serious physical handicap. When east in 1870, Henry Henshaw (1919) made a chance field encounter with this naturalist. "I heard the report of a collecting gun, and investigation soon revealed a young fellow slowly limping his way through the trees, gun in one hand and cane in the other. . . . He came to be an excellent naturalist and a successful and indefatigable bird collector despite his disability, which would have proved an insuperable obstacle to anyone possessed of more than ordinary courage and enthusiasm."

Evidence of the second occurrence of brightly colored Elegant Trogons north of the border was noted in Scott's (1886a) records in 1884. Not witnessing the exciting event himself, he wrote: "Trogon ——?—A species of Trogon undoubtedly occurs casually in the Catalina Mountains. A laborer who had manifested considerable interest in my collections, described to me a bird he had seen only a few hours before, which he believed 'was kind of bird of paradise'" with "'a very long tail, and was bright pink on the breast.'"

A keen observer, Scott (1886b) also noted the local seasonal altitudinal movements characteristic of many birds in the mountainous Southwest. "The migrations here must be considered as occurring regularly in two ways, primarily a north and south migration, and secondarily, though of almost as much import as the other, a vertical migration." Scott (1887) also commented about the erratic or nomadic flight patterns of Red Crossbills by noting that although they were not present in the Santa Catalina Mountains in 1884, the following year they were "abundant and generally distributed throughout the pine woods."

Scott's (1886b) observations while traveling along the San Pedro River to its junction with the Gila included large, white Wood Storks that were "rather common," particularly in the later months of summer. Writing about raptors, he stated that the Red-tailed Hawk "breeds abundantly" and the Crested Caracara was "rather common about Tucson." He also observed Golden Eagles "carrying material for nest building" in the Santa Catalina Mountains.

Following six years behind Scott was Samuel N. Rhoads (1892), who, after a stop in Corpus Christi, Texas, visited Tucson and the Santa Catalina Mountain area for about a month, compiling an impressive list of 126 species. He thought that the California Condor should be entitled to a "place in the avifauna of Arizona" on the basis that a "ranchman" from nearby Oracle "shot a 'Condor' several years ago, near the summit of Mount Lemon . . . to test the range of his rifle."

AN ACCIDENTAL, 1883–1884

The Worthen's Sparrow (*Spizella wortheni*) was secured only once in the United States and that was by Charles H. Marsh near Silver City, New Mexico, in 1884. Considered by the AOU to be "accidental" north of the border, this plain-breasted sparrow is known to nest only in northeastern Mexico. Robert Ridgway (1884) dedicated the sparrow to Charles Kimball Worthen (1850–1909), "who has by his personal efforts done much to develope the ornithology of New Mexico" while working as an illustrator for the Wheeler Survey.

NURSED BACK TO HEALTH, 1883–1900S

After considerable experience in the West, followed by an extensive period in Alaska, Edward William Nelson (1855–1934) returned to Washington, where he, unfortunately, contracted pneumonia and then tuberculosis. His condition worsened until his mother, a Civil War nurse, took him to the White Mountains in Arizona Territory to convalesce under the rather primitive conditions of tent living.

In 1883, after several months recuperating, Nelson joined his close friend Henry Henshaw for a three-month period observing birds on the upper Pecos River in New Mexico. The two men greatly enjoyed the experience since the nesting season was concluding and the fall migration was just beginning. From their camp at seventy-eight hundred feet, they took note of "the assemblage of individual birds into flocks, either of one or of many species" before their movement south. The smaller passerines "all trooped through the forest together, and where one moment reigned perfect silence, the next was enlivened by a chorus of chirps and call-notes, the signals by which the motley throng is held together in an ever moving but united band" (Henshaw 1885, 1886).

During their rambles through the forest, they developed a "friendly rivalry as to which would secure the rarest specimens. . . . This was all done with such genial good humor that it was most enjoyable." It was following one pleasurable incident that Nelson (1932) wrote:

> *He [Henshaw] was a tireless tramper of the mountain sides and on one occasion perpetrated a little practical joke on me that I still recall with amusement. [Northern Pygmy-Owls] were not uncommon but kept so well concealed, high up in big yellow pine trees, that for a long time we were unable to get, or even see one. As I was passing a little grassy park one day I heard the mellow call [Henshaw mimicking] of one of the little fellows apparently low in the branches of a big pine across the opening. Crossing cautiously I slowly circled the tree several times until my neck became painful from the strain of looking upward, while the tantalizing calls came at regular intervals. Finally when I glanced down to be sure of my footing I caught sight of a heel disappearing behind the trunk of the tree. A shout brought forth Henshaw wearing a broad grin of enjoyment in which I joined for I had been completely deceived.*

During the next year Nelson spent several months in Tucson and Gardner Canyon on the east side of the Santa Rita Mountains. At this last location, he apparently camped below a cave used by the Apaches when they trekked from the White Mountains to raid Mexico, but "fortunately, he was there between raids" (Bailey 1923).

In 1890, Nelson joined the Bureau of the Biological Survey where he remained for thirty-nine years. Much of his work was in the West, including fourteen years in California and Mexico. In 1891, Nelson met and hired young Edward A. Goldman (1873–1946) as teamster and camp man, after stopping by the Goldman ranch in central California for buckboard repairs. From this association the two men developed a lasting friendship and an unrivaled working relationship that remained throughout their careers (Goldman 1935).

For much of the next several years, Nelson and Goldman conducted extensive explorations in Mexico for the Biological Survey. They collected the Buff-collared Nightjar *(Caprimulgus ridgwayi)*, which had extended northward into southwestern New Mexico and southeastern Arizona, during this period.

In 1897, Nelson described this nocturnal insectivore in honor of Robert Ridgway for his assistance at the National Museum. They also collected a second bird, the Cordilleran Flycatcher, at the same time in southern Mexico, but it was considered as conspecific with the Pacific-slope Flycatcher until 1989.

In March 1905, Nelson, with Goldman as his assistant, started a year-long survey of the biological resources of the Baja Peninsula. *Lower California and Its Natural Resources* (1966) was the end result of the most exhaustive survey ever undertaken on this 800-mile-long peninsula, which varies in width from 30 to 145 miles. "[W]e traversed the entire length of the peninsula and crossed it eight times from coast to coast. . . . the work was all done with horses and pack animals and included more than 2,000 miles of travel in the saddle." From this survey, Nelson recognized five subordinate faunal districts for the peninsula. Within the districts, he then implemented the life zone concept after C. Hart Merriam (1890), in which he described respective flora and fauna.

The highest mountains occur within the northern region of this lengthy isthmus where, along a short ten-thousand-foot granitic ridge, Picacho del Diablo (Devil's Peak) is the highest point. The western drainage from these steep slopes rushes to the Pacific Ocean. On the east flank, the flow is into the Gulf of California, which separates the peninsula from the Mexican mainland. Many bays, coves, and lagoons indent both coastlines and lying offshore are many picturesque desert islands.

Nelson was the successor of Henry Henshaw as the third chief of the Biological Survey from 1916–1927. He also assisted in the negotiations of the Migratory Bird Treaty between the United States and Canada, which became an Act of Congress in 1918. Goldman played a prominent roll in the enactment of a similar treaty with Mexico, the Convention for the Preservation of Migratory Birds and Game Mammals, in 1936.

RARE SENSITIVITIES, 1883–1900S

ALTHOUGH HE ARRIVED in Tucson about the time Capt. Charles Bendire was posted at Camp Lowell, Herbert Brown (1848–1913) did not became a serious bird collector until after meeting naturalist Edward Nelson (1913a and b) ten years later in 1883. It was through this acquaintance that Brown "expressed the greatest pleasure to have the opportunity to learn something about birds." Up to this time, his prospecting ventures into the mountains had resulted in several "narrow escapes" from Apaches, along with many other life-

*Herbert Brown (1848–1913)
looking into an Elf Owl's nest.
(From* Bird-Lore, *1915)*

threatening situations. By 1890 Brown had a "fine collection," for it was noted and admired by Henry K. Coale (1858–1926) on a "flying trip" from Chicago to several southwestern military posts, including Fort Lowell (Coale 1894).

One of the first accounts on birds written by Brown (1899a) was a report of three men killing a California Condor at Pierce's Ferry near Grand Wash Cliffs in northwestern Arizona. "The men had no rule, so measured it with a gun. It was over a gun length in height and more than three gun lengths in the spread of its wings."

Brown (1901b), a naturalist with considerable experience with nesting Bendire's Thrashers, wrote of the "delightful" characteristics of this seldom-heard songster. "Once, and only once, I heard one in a grand outburst of song. I had to positively convince myself that I was not mistaken, and I was not. I then

THE DWINDLING FRONTIER

realized that if unsung melodies were sweet, this feathered grace would queen the plains." At the turn of the century, M. French Gilman (1871–1944), a naturalist who had also reveled in the songs of desert thrashers, agreed with Brown. "As for singing, the Bendire has them all beaten. The others are fine singers indeed, but their repertoire is limited. . . . No two seem to sing exactly alike and some of the songs are quite distinct from others" (Gilman 1909).

After correspondence with Bendire, Brown (1901a) concluded that he had found, in a certain cottonwood tree in the foothills of the Rincon Mountains, the same nest of a Zone-tailed Hawk discovered by Bendire fourteen years earlier. Presumably, this was the same location at which Bendire (1892) had his thrilling encounter with Apaches.

While in Sonora, Mexico, Brown (1904) became the first naturalist to hear the "startling and unexpected" call of the "Masked" Northern Bobwhite. In 1883, he collected this handsome quail in Arizona. Extirpated before the turn of the century, in 1883 this quail was restricted in the United States to a very limited grassland area that extended only fifty miles north of the border along a hundred-mile strip east of the Baboquivari Mountains. As a resident during the devastating drought of 1892–1893, Brown (1900, 1904) attributed the demise of this quail, along with the "obliteration of bird life in the so-called desert portions of the Territory," to the "overstocking of the country with cattle, supplemented by several rainless years." William Brewster (1887) also received specimens taken about fifty miles south of the border in Sonora by John C. Cahoon (1863–1891) who, in another adventure, fell to his death while collecting.

Brown recorded two unusual species for southeastern Arizona that have never been duplicated (1899c, 1906). They are the Scarlet Ibis and the Anhinga. He is also credited with collecting the first Streak-backed Oriole and Scarlet Tanager in the region (Phillips, Marshall, and Monson 1964).

Obviously a man of rare sensitivities for the period, Brown (1903) recorded the events following a rainy and windy night in the winter of 1899. Dotting the surface of the Santa Cruz River outside Tucson, he noted "bunches of ducks" which, once discovered by the local men and boys, "were shot at without mercy, decency or common sense, and although it was tails up at the flash of a gun they were eventually tired down and killed off."

On a more positive note, Brown (1903) observed the "coming and going" of thousands of American White Pelicans near Yuma during the spring of 1902. "I did not think it was possible to see so many pelicans together. They occupied a wide sweep of sand, left by an overflow of the Colorado and, at a distance, resembled great banks of snow."

A year before Brown's observation, settlers arrived in nearby Imperial Valley and began the first attempts at diverting Colorado River water, in this case through a simple wooden headgate and canal, to the farmlands in southeastern California. By 1905, after a series of floods, the headgate and canal gave way and the Colorado River channeled out of control into the Salton Sink. In 1908, Joseph Grinnell (1877–1939), whose name echoes in the halls of the University of California at Berkeley, cruised the rapidly rising Salton Sea in an effort to determine the waterbirds nesting in the area. Aboard a small boat christened the "Vinegaroon," Grinnell and his companions voyaged from Mecca, at the north end of the Salton Sea, forty miles southeast to Echo Island, where he discovered the "southernmost recorded nesting-colony of the American White Pelican" (1908).

Ironically, the island, which was formed by the hand of man, was eventually engulfed by his actions, ending its potential as a bird-breeding site. This intervention on the wild free-flowing characteristics of the Colorado River also marked the beginning of a series of projects that ultimately led to the complete control and destruction of much of its diverse riparian qualities.

Brown organized and was the first curator of the University of Arizona's museum, and his bird collection was its nucleus. His life experiences were quite varied: he served as Superintendent (1898–1902) of the Yuma Territorial Prison (where an Apache convict helping him was bitten by a rattlesnake [1899b]), a reporter, a Tucson newspaper editor, and clerk of the Superior Court of Pima County.

VISITORS FROM MEXICO, 1885–1900S

THE YEAR 1885 in southeast Arizona Territory was a period of great uncertainty, as indicated by the local Willcox newspaper, the *Southwestern Stockman*. On June 6, the paper reported that "Col. Mike Gray came up from his ranch at old Camp Rucker last Saturday, but news of the Indian outbreak, which he then first heard, hastened his return home before he had time to visit with his many friends in this vicinity."

Then, about three months later on September 26, the same paper reported an unusual event that was not only exciting to the observers, but to science—a flight of unexpected parrots. "During the past summer a flock of several hundred parrots made its appearance in the neighborhood of old Camp Rucker, in the Chiricahuas, and after remaining a few weeks disappeared as mysteriously as it came. They were the first ever seen in those mountains, but the fact of their being there is vouched for by Mike Gray and family." Other than a specimen by J. W. Audubon (Baird, Cassin, and Lawrence 1858; Sclater 1857c), this was the first reported occurrence north of the border of the Thick-billed Parrot, a colorful bird with brilliant green feathers and scarlet on the bend of the wings and forehead.

Apache Indians probably seldom, if ever, saw these parrots north of the border. However, when venturing into Mexico, they saw caged birds on occasion that they believed were mocking them. Their superstitious belief was that, "If a bird talks, there must be a witch in him" (Opler 1941).

The first naturalist to observe the Thick-billed Parrot north of the border was Richard D. Lusk (1900) fifteen years later. His interesting comments about the sudden arrival of these Mexican visitors to forage among the pine forests in the Chiricahua Mountains reveal one of the reasons for their rapid decrease in numbers. "[T]here came . . . a flight of nine or ten parrots, scolding and chattering and calling in a language which was neither English nor Spanish, but may have been some Indian tongue, or, indeed, that of the old Aztecs of Mexico themselves." After arriving near Lusk's camp, "they tarried—many of them I regret to say, for aye, for the timbermen in a polecutter's camp hard by, carried away by the novelty of the visitors, began slaughtering them. . . . And I, of course, must have a couple of specimens of the rare straggler (?)."

Four years later another flock of "700 to 1000" parrots appeared in "Bonita Park" at the north end of the Chiricahuas, where they "greatly excited the miners, who were inclined to consider it a lucky sign, with 'strikes' sure to follow" (Smith 1907). This event and several others were also shared by the Hands brothers, Frank Henry (1863–1936) and E. John (1866–1939), ranchers, miners, and owners of the Hilltop Lead-Silver Mines in the northern part of the Chiricahuas. Another brother, Alfred, was slain by Apaches in Cave Creek in 1896. Their serious interest in birds developed after an acquaintance with Californian John E. Law (1877–1931), who, with failing health, was attempting to do a full accounting on the Chiricahua birds.

In an unpublished letter to Law regarding the events of January 1918, Frank Hands described the local boys who were after the parrots "on shares" and wrote that they were "practically camped on their trail" during the hunt.

His letter also contained amusing observations on feeding parrots. "They would fly to a snow covered limb, turn over, and grab the underside with their feet, woodpecker fashion, pulling themselves along with their bill after the acorns, and occasionally dropping into the snow after those that fell. Wading in the snow with their short legs, and solemn appearance, was very ludicrous, and gave us several laughs. The poor beggars were having such tough sledding, that I hadnt the heart to kill them" (Law 1913–1919).

Some years later, two distinguished naturalists carefully assembled the sight records of parrots in Arizona (Vorhies 1934; Wetmore 1935). Wilmot W. Brown, collector for John E. Thayer of Massachusetts, probably secured the first eggs of this species in Chihuahua, Mexico (Thayer 1906). William H. Bergtold (1865–1936) also worked with them in the same area (Bergtold 1906).

SEAFARING NATURALISTS, 1881–1900S

𝕄OST OF THE early naturalists were either exposed to or had knowledge of ornithologists, who greatly influenced their study of birds. For Lyman Belding (1829–1917), this exposure did not occur until, at the age of forty-seven, he gained from James G. Cooper's *Ornithology* (1870) a new and fuller interest in birds. Then, after sending a number of specimens to Robert Ridgway and Spencer Baird, Belding (1900) gratefully received "many kind attentions" by which his "zeal for the work was greatly stimulated." Complimenting their actions, he continued, "I do not think this kind of encouragement was exceptional, for I think Profs. Baird and Ridgway were always glad to assist the student of natural history."

Baird and Ridgway asked Belding to visit Guadalupe Island in 1881. Unable to carry out their wishes because of the island's remoteness and lack of inhabitants, Belding's attention turned to the west coast of Baja California. He landed on the largest of the Mexican coastal islands, Cedros, a promontory that rises almost four thousand feet above the sea at latitude 29°15′. Although it contains an isolated forest of Bishop and Monterey Pines at the upper elevations, less than a score of bird species occur here. He found it "quite destitute of birds" but did manage to collect a Double-crested Cormorant.

Belding (1883a and b) collected for two winter seasons in the Cape Region. While among the mangrove thickets of the lagoons, he secured the elusive Clapper Rail. He also obtained the first nest and eggs of the Costa's Hummingbird at La Paz. Landing at Bahía Quentin, Belding secured a dark resident form of the Savannah Sparrow which, in this coastal area, belongs to the *"beld-ingi"* group. The northern Mexican islands, Isla Coronado, produced for him the large shorebird, the red-billed American Oystercatcher.

The geographic isolation of the peninsula, especially the southern Cape Region, illustrates the potential for evolving speciation. Belding's Yellowthroat *(Geothlypis beldingi),* named in honor of the naturalist by Robert Ridgway in 1882, is an allopatric species that has become morphologically distinct from the mainland Common Yellowthroat.

In noting his discovery, Belding (1900) commented that "I have often wondered why the sharp-eyed, indefatigable Xántus did not see *Geothlypis beld-ingi* on the San Jose river, where he spent a great deal of time, and also if it had rapidly changed since he was there some thirty [twenty] years before." Belding was among several succeeding ornithologists who questioned the field activities of this earlier naturalist.

Belding also wrote several major reports on birds of central California. Walter K. Fisher (1918), a western field naturalist for the Biological Survey, took note of his work. "Mr. Belding was a painstaking and accurate observer, a conscientious recorder, and had in fact the real spirit of research. He hated inaccuracy and exaggeration. What he did he did well, and his limitations were those imposed by his isolation and lack of early training in scientific pursuits." Albert K. Fisher (1920), another Biological Survey naturalist, noting Belding's "genial" nature, added that "there is little wonder that he was so popular and so eagerly sought after by old and young, especially when found in the outing season in his favorite haunts in the Sierras."

Several naturalists are remembered for their often dangerous voyages in search of new discoveries in the coastal waters of Baja and southern California. Certainly Walter E. Bryant (1861–1905) is among that special group. Born in California, he "inherited a love of nature and a love of adventure," which he continually developed until he died, shortly after contracting a fever on a business trip to San Blas, Mexico (W. K. Fisher 1905). In 1886, Bryant became curator of birds at the California Academy of Sciences following his second trip to Isla Guadalupe.

In writing of the voyage, Bryant (1887) described the remote location as "almost unknown" with "charts quite unattainable." Like Edward Palmer, who ten years earlier had an unexpected prolonged stay, Bryant fell into a similar circumstance after planning only a six-week trip. "The accommodations, moreover, were not the most suitable, nor were the comforts of life in excess of the demand for them. As a result of three and a half months' sojourn on the island, the number of known species has been increased by twenty-seven, making a total of thirty-six known to the island."

In 1887, Bryant, recognizing morphologic differences between the Leach's Storm-Petrel and a new petrel he had collected on the island, applied a subspecific name to the latter. Its common name, incidentally, came into acceptance after the Dutch ornithologist Coenraad Temminck named it in honor of English naturalist William Elford Leach (1790–1836). Bryant's subspecific description and the specimens he collected eventually became known as the Guadalupe Storm-Petrel.

Since both petrels breed on this island, Bryant's fascinating notes about them returning to their nesting burrows during the last night of a three-day gale could have been about either or both.

> *No ornithological work was possible, and nothing could be done . . . but to hug the camp fire. At midnight . . . my companion awakened me to announce that some "little owls" were flying about. Every few minutes a bird would pass the small circle of light or hover for an instant in the flow above the fire, while from the enveloping darkness their call and replies could be clearly heard . . . their movements, with flight as erratic as that of a bat. The birds came about my camp only on the darkest nights or, if any were flying during moonlight, they were entirely silent. After the setting of the moon, however, even though as late as four o'clock in the morning, they would make their appearance with their peculiar call.*

Unknown to Bryant at the time, these pelagic storm-petrels, or "sea-patters," locate their burrow by smell, through olfactory nerves in their prominent fused tube, or nostril, on top of their bill. Known to have bred only on Guadalupe Island, the Guadalupe Storm-Petrel is now extinct, probably due to the introduction of feral animals. Wilmot W. Brown, collecting for John E. Thayer, recorded his later visit to the island. "The mortality among these birds from the depredations of the cats that overrun the island is appalling—wings and feathers lie scattered in every direction around the burrows along the top of the pine ridge." He estimated between six and eight thousand goats, turned loose many years ago, were "responsible for the destruction of its flora" (Thayer and Bangs 1908).

John Thayer (1862–1933) deserves some note in that he was a sponsor of several collectors who visited the Mexican interior and its coastal waters (Thayer and Bangs 1907). As a prominent amateur ornithologist, he established the Thayer Museum of Ornithology, which consisted of a huge collection that was eventually donated to Harvard (Phillips 1934). Thayer's Gull (*Larus thayeri*), which reaches its southern winter limits along the Baja California coast, was named for him by Harvard ornithologist W. S. Brooks in 1915.

Turning again to the land birds of the Guadalupe Island, Bryant (1887) observed Rock Wrens, the most numerous of the island birds, carrying "flat pebbles ranging in size from a Lima bean to a half dollar," which they used to reduce their rocky nest site entrances.

One of Bryant's major contributions was "A Catalogue of the Birds of Lower California" (1889), in which he supplemented his notes with those of Lyman Belding and Alfred W. Anthony to produce an annotated list of 320 species. Bryant (1891b) also provided many insights into the local culture of the Cape Region. Joseph Grinnell (1905) recognized him as a major contributor to Pacific coast ornithology. "The life-histories of many of our remotely restricted species, would remain to-day almost wholly unknown, if Bryant had not spent lonely months in their study, and then composed what he learned in the form in which we find it now so instructive."

Another naturalist, Alfred W. Anthony (1865–1939), a mining engineer by profession, added much to our knowledge about ornithology because of his travels in New Mexico, Arizona, and Baja California and its many offshore islands. Most of his bird specimens, numbering ten thousand, were sent to the Carnegie Museum in Pittsburgh. Before traveling to the west coast where he accomplished his most notable work, Anthony (1892) was engaged in mining activities in the extreme southwestern corner of New Mexico Territory from 1886 to 1889. He recorded 129 bird species in the area, but lamented about that experience: "Owing to hostile Apaches it was necessary to avoid the higher mountain ranges—the Hachita and Animas, as well as favorable points in the Sierra Nevada just south of the boundary. Many interesting records were thus lost."

In addition to being one of the first naturalists to visit the remote San Pedro Martir in northern Baja California, Anthony had a special interest in island fauna. His fascination for seabirds grew, especially after he purchased a schooner, which enabled him to access the remote islands of Baja and southern California. Information about the life histories of many pelagic birds, so difficult to obtain, was added through his efforts. Several of his papers deserve more than a cursory review of titles. "The Fulmars of Southern California" (1895) provides interesting comparisons between the occurrence, abundance, and habits of Northern Fulmar, Black-footed and Short-tailed Albatrosses, and Sooty, Pink-footed, and Black-vented Shearwaters. Anthony (1900a) also found Black-vented Shearwaters nesting on several Mexican islands.

His search for nesting colonies of Black Storm-Petrels resulted from his many enjoyable earlier experiences of observing their genus performing erratic, but dainty, flight behavior over the ocean surface (Anthony 1898). "From the day that I saw my first Petrel dancing over the waves of the Pacific none of the birds of southern California so thoroughly interested me or so completely baffled all attempts at a more intimate acquaintance."

Anthony expressed great interest in the habits and behavior of many seabirds. Their winter arrivals and departures to San Geronimo or Natividad Islands were typical, as Anthony (1906) observed: "still thirty miles or more distant, long, straggling flocks of cormorants, loose scattered companies of gulls, and small military squads of . . . Brown Pelicans, all converging toward one point. As the island grows larger and the sun sinks lower birds become more and more plenty, flocks hurry by with greater frequency and with an air of business." The three species of cormorants he observed were Double-crested, Pelagic, and Brandt's. The latter cormorant bears the name of German ornithologist Johann Friedrich von Brandt (1802–1879), who was appointed to the Zoological Museum at the Academy of Sciences in St. Petersburg.

Anthony often spent nights at sea with birds, thereby enjoying their company and learning their habits. Once, on a trip from San Diego to North Coronado Island, Mexico, Anthony (1899) secured a fifteen-foot skiff and "at dark I started—alone,—because, as some one said, no one was fool enough to go with me and at night . . . with nearly a full moon, it was altogether the most enjoyable time for the twenty-two mile pull to the islands." The floating kelp beds were perching places for gulls, terns, and cormorants, while "flocks of six or eight pelicans would pass like grey ghosts in the moonlight, flying in 'pelican order,' each just behind and a little to one side of the one preceding." All night long there was the "going and coming" of Cassin's Auklets and Xantus' Murrelets, each making "perhaps several trips each night to satisfy the cravings of the ever-hungry squabs."

As an inquisitive naturalist, Anthony understood that while the diurnal activities of some birds cease, others begin afresh with their nocturnal routine. Many times while going ashore at night to investigate breeding pelagic bird colonies, he would naturally observe other species. On San Benito Island, Anthony (1900b) was greeted on his landing by the "rattling alarm" of the American and Black Oystercatchers, which overlap their ranges in these southern coastal waters.

In 1896, Horace A. Gaylord (1897) accompanied Anthony on an "interesting sojourn" to a most "out-of-the-way-place," Guadalupe Island. His notes indicate that some of "the birds peculiar to the island are so tame that some will occasionally attempt to alight on the barrel of the gun aimed for their destruction." During their four-day stay, they endeavored to take one of the only "three or four" remaining Guadalupe Caracaras.

Anthony (1924) expressed great concern about plume hunting, not only in Mexican waters, but for pelagic birds that breed on many distant islands and wander throughout the world. Three species of albatross that formerly ranged regularly into western Mexican waters were greatly reduced during his time. The Black-footed and Laysan Albatross were plundered for feathers on islands between Hawaii and Japan in the early 1900s (Dutcher 1904). Also, the Short-tailed Albatross was all but exterminated on another Japanese island. Almost a century later, the Laysan Albatross has been staging a significant comeback and has extended a breeding colony to the Guadalupe Islands, the first known nesting record east of the Hawaiian Islands (Dunlap 1988).

Vernon Bailey (1941) considered Anthony to be one of "our great naturalists" who "left an enviable record of achievement," but it was not without hardships. In 1899, while sailing south thirty miles north of Magdalena Bay along the west Baja California coast, Anthony's schooner, *Stella Erland,* was totally wrecked.

H. B. Kaeding, who sometimes assisted or collected for Anthony, also made many interesting Mexican island observations. On a trip down the coast of Baja California in 1897, Kaeding (1905) listed 159 birds, among which was a common migratory species with an obscure name, the Western Sandpiper *(Calidris mauri).* In 1839, Charles Bonaparte listed this sandpiper with no explanation of its application in honor of his Italian botanist friend Ernesto Mauri (1791–1836). Eighteen years later, Jean Cabanis, who is credited with describing the bird, applied the eponym only after a series of naming complications and some persuasion by Bonaparte (Palmer 1931).

At the close of the century, Spencer Baird appointed Charles Haskins Townsend (1859–1944) to the U.S. Fish Commission (Palmer 1947a). From 1889 to 1899, and again in 1911, Townsend (1890, 1923, 1927) cruised the waters of Baja and southern California. It was from the *Albatross,* a research ship designed by Baird, that Townsend collected the Townsend's Shearwater from among the Revillagigedo Islands some four hundred miles southwest of the tip of Baja California.

Several naturalists contributed to the ornithological history of the Channel Islands off the southern California coast. Some of their work expanded to island speciation relative to mainland ancestry. They were N. S. Goss (1884), E. W. Blake, Jr. (1887), C. P. Streator (1888), R. H. Beck (1899a and b), J. Mailliard (1899, 1900), and G. F. Breninger (1904). Joseph Grinnell (1902a) was just beginning his life-consuming work on California birds.

A growing disparity of bird observations was already developing between earlier naturalists, such as J. G. Cooper, who visited Santa Catalina Island in the 1860s, and a later group arriving at the close of the century. In the case of raptors, Cooper ([1870] 1974) reported "more than thirty . . . [Bald Eagles] in young plumage . . . and their nests were numerous among the inaccessible cliffs." The succeeding naturalists also indicated that human pressure marked the beginning of a decline and eventual extirpation of the Bald Eagle, Osprey, and Peregrine Falcon on this and several other Channel Islands.

The nearby mainland was also undergoing human change, which was destined to substantially impact birds. The solitude of sandy beaches and quiet lagoons that were the breeding areas of Snowy Plover and Least Terns were among those to be dramatically affected (McCormick 1899). H. A. Gaylord (1899) rendered a descriptive account of the "ever-changing representation of bird life" during a spring migration in the San Gabriel Valley reminiscent of past avifauna splendor.

LIFE ZONES AND A LADY BRONCO RIDER, 1889–1900S

IN 1872, CAPT. J. G. Bourke ([1891] 1971), accompanying Gen. G. Crook's campaign against the Western Apache bands on the Mogollon Plateau, wrote the "rapacious forces of commerce" against the "forest primeval" was about to be unleashed by the establishment of the Ayers-Riordan sawmill at Flagstaff. To Bourke, the destruction of this Ponderosa forest was "scarcely inferior" to "man's inhumanity to man" for the trees were "nearly human: they used to console man with their oracles."

Seventeen years later in the same beautiful area, C. Hart Merriam (1855–1942), intrigued with the premise that climatic conditions affect the altitudinal succession of mountain biota, set off for Arizona to collect data to support his hypothesis on the steep slopes of the San Francisco Peaks (12,633 feet). Rising over five thousand feet above the seven-thousand-foot Colorado Plateau, these volcanic mountains provide an assortment of temperature and moisture variables that result in vegetative zones rich in diverse fauna.

Using an aneroid barometer to establish elevations, a clinometer to determine slope, and a compass for orientation, Merriam (1890) developed his "life zone" concept, which he reported in the "Results of a Biological Survey of the San Francisco Mountain Region and Desert of the Little Colorado in Arizona." After general acceptance, his terminology for the zones from the summit down to the base were Arctic-Alpine, Hudsonian, Canadian, Transition, and Upper and Lower Sonoran. An original thinker and a contributor to the developing science of ecology, Merriam believed that from his supposition, a much broader view could be extrapolated of the distribution of plants and animals across North America (Sterling 1974). In his thesis, he listed characteristic birds that could serve as indicators of their respective "life zones" during the breeding season (Monson 1964). During the same period, the theories of Joel A. Allen (1893) were also embracing geographical origins of broad faunal areas which included birds.

During August 1889, Merriam, his wife Elizabeth, and botanist F. H. Knowlton joined with Vernon Bailey, his assistant, on a second ascent to the summit. "We saw two golden eagles there at about 12,000 feet. . . . A pair of peregrine falcons seem to have their headquarters in some high cliffs in the crater. . . . We ate broiled eagle for supper" (Phillips, House, and Phillips 1989). Leonhard Stejneger, recovering from a serious lung ailment, took leave from the National Museum to participate in the expedition.

Merriam was a founder of the AOU and later served as its president from 1900 to 1903. He was selected to be the first chief of the U.S. Office of Economic Mammalogy and Ornithology, which later became the Biological Survey (Palmer 1954). A cinder crater east of the San Francisco Peaks is named in his honor. Among Merriam's (1916) many friends was an unconventional camper, but an inspiring preservation advocate, John Muir (1838–1914).

In 1889, the extensive work of Vernon Bailey (1864–1942) with the Biological Survey began under Merriam and continued intermittently for the next forty-six years. Primarily a mammalogist, he was also an excellent ornithologist and later became chief naturalist for the Survey (Palmer 1947b). In recogni-

Florence Merriam Bailey (1863–1948). (From Bird-Lore, *1916)*

tion of his work in Texas, Vernon Bailey Peak (6,670 feet), four miles north of Emory Peak in the Chisos Mountains of the Big Bend country, was named in his honor.

In 1899, Bailey married Florence (1863–1948), the sister of Merriam, who became the most well-known woman ornithologist of that period. Several years earlier, she may have contracted tuberculosis from her mother, but her health was fully regained after traveling west to the climates of California and Arizona (Kofalk 1989). It was following her visits to Twin Oaks, a Merriam homestead in southern California, that she introduced a popular approach to observing birds in *A-Birding on a Bronco* (1896a). After returning East and getting married, Florence Merriam Bailey again traveled to the Southwest with her husband, where

The Dwindling Frontier

she wrote about birds for several publications, including the *Auk, Bird-Lore,* and the *Condor.*

She also wrote a widely accepted book, *Handbook of Birds of the Western United States* (1902), which provided Olaus J. Murie (1889–1963) with some insight into her views on bird collecting. He wrote: "Revering science with a deep devotion, and with skilled firsthand experience, she still saw more in a specimen than a skin" (Oehser 1952). Henry Henshaw (1920a) added that although she was "much interested in her brother's collections, her personal predilections were for the study of the living bird over the stuffed specimen, and she was ever alert to urge the advantages of this method of study as against the less humane use of the shot-gun."

With an understanding of the issues, she also saw fit to add in her book the defensive but poignant thoughts of her scientist collector husband. "Naturalist collectors are far from being the ruthless destroyers of life they are often supposed to be. It is, indeed, those who collect the birds, study them most deeply, and know them best, who are doing the most for their protection" (Bailey 1902).

She was also among the first to include in her book, as did Joseph Grinnell (1902a), a section on Merriam's life zone concept. Working vigorously for bird protection and education with the Audubon Society was also a high priority for her. She, therefore, thought it was appropriate for T. S. Palmer to contribute a section on the "potent influence" the Lacey Act of 1900 had on prohibiting the interstate commerce of birds. Irene Grosvenor Wheelock, writing her fine book *Birds of California* (1904), referenced the work of her contemporary, Florence M. Bailey.

After succeeding C. Hart Merriam and Henry W. Henshaw, the third chief of the Biological Survey, Edward W. Nelson, assigned Florence M. Bailey the task of expanding some of the bird data on the distribution and migratory records of the then late Wells W. Cooke (1858–1916) into the first comprehensive book on southwestern avifauna (Palmer 1917b). Bailey (1928), already a superb perceptive field naturalist, displayed her exhaustive thoroughness as a researcher in this work, which was entitled *Birds of New Mexico.*

Along with compiling a tremendous amount of ornithological data in this volume, Bailey also displayed a wonderful writing style in which she expressed her appreciation for birds and demonstrated her vast field experience. Her description of the lovely, early-evening, flutelike vocalizations of

Hermit Thrushes in the spruce forests of New Mexico exhibits her extraordinary talent: "[F]rom unseen choristers a serene uplifted chant arose, growing till it seemed to fill the remote aisles of the forest. Sometimes a silvery voice would come from the open edge of the dark forest, where the singer looked far down the mountainside and out over the wide mesa-clad plains."

Paul H. Oehser (1952), in his memoriam of Bailey, wrote that "As an ornithologist. . . . Her forte was . . . observing and describing what she saw afield. . . . [Her] conscientious awareness of avian literature . . . was anything but meager, yet she never carried it ostentatiously."

In addition to her knowledge about birds, Bailey also had a strong determination to endure the rigors and dangers of a field naturalist. Frank Chapman (1916) commented in tribute to her that "no woman has studied the wild birds of America so systematically, so thoroughly, and so carefully as she. The amount of field-work she has done is perfectly astonishing, and probably few women have spent so many days in the wilds, or so many nights under canvas."

EMERGING ARTIST-NATURALISTS, 1893–1900S

THE PORTRAYAL OF birds with pencil and brush has often brought to our lives an appreciation and enjoyment of the beauties of nature. And, perhaps it should be said, the skills and techniques of a true bird artist resonate his personal involvement and understanding of his subject. His comprehension as an artist-naturalist, with his field or museum experience, brings another dimension to his craft. These fundamental essentials embraced by John J. Audubon, Andrew Grayson, and Robert Ridgway certainly reflect their commitment.

Roger Tory Peterson (1908–1996), who has contributed more toward the appreciation of birds than perhaps even Audubon, wrote of the intricacies in attaining perfection with bird art. "To produce a drawing that is ornithologically correct is difficult enough, but to add to this the intangible assets of good draftsmanship and good painting is something few ever achieve" (Peterson 1942).

The turn of the century brought two artists of considerable talent to the borderland region. The first was Louis Agassiz Fuertes (1874–1927), who produced such illuminating art that many considered him to be "America's greatest bird painter." A few books that display his remarkable pre-eminence include *Key to North American Birds* (1872) by E. Coues, *Handbook of Birds of the West-*

Montezuma Quail, Cyrtonyx montezumae *(Vigors),
1830. Painted by Louis Agassiz Fuertes, ca. 1903. (Courtesy
of the U.S. Fish and Wildlife Natural History Art Collection, Academy of Natural Sciences of Philadelphia,
Ewell Sale Stewart Library)*

ern United States (1902) by F. M. Bailey, and *The Bird Life of Texas* (1974) by H.
Oberholser. Some of the journals that recognized his talents and had his artwork grace their pages were the *Osprey* and *Bird-Lore*. His cover design for the
Auk has remained since 1915 (Peck 1982). Many of his paintings are displayed in
the Academy of Natural Sciences in Philadelphia.

In 1900, as a member of the Biological Survey party for five months in
New Mexico Territory and west Texas, Fuertes (1903) unveiled his little-known
but exceptional writing talents as a field naturalist. "Of all the bizarre and
curious creatures that live in our county, it would be hard to find one more
arbitrarily marked, or colored more apparently in opposition to the laws of pro-

tective gradation and coloration than the Mearns [Montezuma] quail . . . when we see the fantastic little cock Massena with his dark chestnut breast, jet black belly and flanks, and harlequin-painted head, it is hard to conceive how he was ever able to qualify in the race for survival among a group of birds so marvelously protected as his congeners."

During the same period, Harry Oberholser often told of how, while in west Texas with Vernon Bailey, they rescued the third member of their party, Fuertes, from a cliff ledge with a rope because of the artist's "impetuous" act of becoming stranded while attempting to secure a Zone-tailed Hawk (Aldrich 1968).

Frank Chapman (1928) remarked about Dr. Coues's "magnetic personality" in guiding and encouraging young Fuertes. "Nor can we value too highly the influence which Coues exerted in developing Fuertes' talents and in shaping his career." Coues elaborated in the *Osprey* that Fuertes was "the most promising young artist of birds now living, and one whose work already places him in the very first rank." It is fitting that Fuertes, with so much to give, took ornithologist and painter George Miksch Sutton (1898–1982) "under his wing" in his early formative years as a tyro artist (Sutton 1979).

Thoughtful words relating to those having sensitive avian values were wonderfully phrased by Chapman (1928) in tribute to Fuertes. "Love of birds as 'the most eloquent expression of nature's beauty, joy and freedom,' is the gift of every one who hears the call of the outdoor world. But that instinctive, inexplicable passion for birds which arouses an uncontrollable desire to know them intimately in their haunts and to make them part of our lives, and which overcomes every obstacle until in a measure, at least, this longing is gratified, is the heritage of the elect; and few have been more endowed than Louis Fuertes."

The second exceptional borderland artist of the period was Major Allan Cyril Brooks (1869–1946). Born in India, Brooks moved with his English parents to Canada, where he remained in residence much of his life. Not only an enthusiastic ornithologist, he was also one of the truly excellent artists of American birds. He visited the Southwest immediately after the turn of the century, painting many borderland species. Through the years he traveled the border areas of all four states adjoining Mexico. Brooks's close association with Fuertes was of such mutual respect that they sometimes painted together in Fuertes' studio (Brooks 1938; Harris 1946).

The four volumes of *Birds of California* (1923), authored by William Leon Dawson (1872–1928), contain many superbly delineated full-page color plates.

Allan Brooks (1869–1946) working on his painting of an immature Golden Eagle, which he titled Missed, *in his studio at Okanagan Landing, British Columbia, 1939. (Courtesy of Elizabeth Brooks)*

Other books in which Brooks's work appears include the *Birds of the Pacific States* (1927) by R. Hoffmann (1870–1932), *Birds of New Mexico* (1928) by F. M. Bailey, and *The Book of Birds* by editors G. Grosvenor and A. Wetmore for the National Geographic Society (1932–1937), which includes 950 color bird portraits. Within the covers of H. Brandt's *Arizona and Its Bird Life* (1951), Brooks mastered action artistry in "Missed," when he depicted a grassy plain with an immature Golden Eagle attempting to capture a dodging jackrabbit. He was also published in several journals.

Dawson (1913) recognized the many talents of Brooks and wrote that he had an "inexhaustible store of exact information as to plumage changes, evanescent colors, scutellation of tarsi, and all else that pertains to the external appearance of birds." This was certainly borne out by Brooks's (1922) query, when he investigated the color differences in the feet of Western Gulls. Brooks stated that the "correct colors of all soft parts is of the primest importance in the Laridae, where so many closely allied species have feet of very different colors." He further referenced the southern form as having "yellow feet" compared to all others having "flesh colored" (Dwight 1919; Ridgway 1887c). In 1982, the Yellow-footed

Gull (*Larus livens*) was finally recognized as a distinct species. Ornithologists determined that the dark-mantled gulls breeding in the Gulf of California do indeed have yellow feet, they have differing vocalizations and nesting habits, and they remain completely separated from those on the western coast of Baja California (McCaskie 1983).

MISFORTUNE AMONG THE PINES, 1899–1901

The "Vale of Vespertina," named by young ornithologist Francis J. Birtwell (1880–1901) and his bride Olivia, was a special place where they sought out many forest birds on a slope behind their cabin in Willis, New Mexico. One bird that held Birtwell's (1901) special interest and for which they attached the sobriquet to the locality was the Evening Grosbeak, described by W. Cooper in 1825. Cooper, perhaps thinking they sang only in the evening, may have mistakenly applied the scientific species name to the bird *vespertinus,* which is derived from the Greek and means "evening." Other naturalists later noted that the birds vocalized from sunrise to sunset.

For most of June 1901, the couple observed the habits of these gregarious birds as they built their nests and began laying eggs. Determining the egg sets to be complete, Birtwell began the slow climb to collect three sets in the towering pines. It was on his third attempt that the unfortunate mishap occurred. Somehow in the dangerous sixty-five-foot ascent, he "became entangled in the rope and strangled in the presence of his bride" (Ewan and Ewan 1981).

Born in England, Birtwell had traveled to the United States in anticipation of continuing his education. Contracting tuberculosis while in the east, he resumed his studies at the Territorial College of Agriculture in New Mexico in 1899. His accepted thesis for graduation was to be "The Ornithology of New Mexico." The notes he recorded were included among those compiled by Miss Fanny Ford (1911) (Mrs. Arthur Sloane) in the first bird list for the state of New Mexico.

During his brief life, Birtwell was fortunate to have become acquainted with Elliott Coues, who had "predicted" for him a "great future." Borrowing some thoughts of those who had known him, Olivia (1901) summed up his accomplishments: "[H]is keen powers of observation, his independence of thought, and his tireless zeal made him a young man whose career promised to be of great service to ornithology."

⤳ 7RRESPECTIVE OF THE rigidly defined borders of New Mexico and Arizona with adjoining Mexico, several pioneer faunal species occur in the region. For several bird species, the borderland represents a marginal and somewhat fluctuating frontier. And, the higher mountain elevations and the lower valleys form biotic communities or faunal limits exhibiting great diversity.

At the close of the century, special attention was given to this sector of the Mexican borderland by a coterie of naturalists who appear in this segment. The first was ornithologist Albert K. Fisher (1865–1948), an AOU founder and later its president from 1914–1917. Throughout his life, he "retained his keen interest in natural history and an alertness of mind that was an inspiration to all who knew him" (Uhler 1951).

While working with the Biological Survey in the border region, Fisher visited the Chiricahua Mountains in 1893 and 1894. Major Charles Bendire (1895) quoted him about his experiences with one of the common nightjars during the last day of May in 1894. In the Southwest, habitat and elevation are among the factors that determine their occurrence. "The Whip-poor-will's note was not heard at Fort Bowie [approx. five thousand feet]. . . . [W]e made camp at the mouth of Rucker Canyon [approx. fifty-three hundred feet], some forty miles south . . . [and] the last day of the month, we heard a few, and a couple days later found the species abundant higher up the same canyon, among the pines [approx. sixty-five hundred feet]."

Climbing to "Fly[s] Park [approx. ninety-two hundred feet] . . . the species was very much less common, though a few were heard every night. . . . [T]hey often alight on a prominent rock or dead stub, from which they launch out after passing insects."

At Flys Park, Fisher (1894), with young Fred H. Fowler, son of Captain Joshua L. Fowler (1846–1899), who was posted at Fort Huachuca, added the first White-eared Hummingbird north of the border. Fowler (1903) also recorded the occasion. "We had not gone fifty yards from the tent, when the Doctor saw, perched on a twig, a hummer which had a decidedly white patch behind its eye. He called my attention to the peculiarity." Richard Lusk (1921) also found the same species in the Huachuca Mountains during that year.

Other naturalists from California were also preparing to explore these intriguing mountains. At the age of fourteen, William "Billy" W. Price (1871–1922), following his father's death, started from Riverside, California, for

Arizona to explore the mountains and deserts of southeastern Arizona Territory. Then, at the age of seventeen, Price (1888) added the first Rose-throated Becard *(Pachyramphus aglaiae)* to our fauna "in the pine forests of the Huachuca Mountains, at an elevation of about 7,500 feet, and seven miles north of the Mexican boundary." In 1839, French ornithologist Baron Frédéric de Lafresnaye dedicated this tropical flycatcher to Aglaé Brelay for her devotion in assisting the work of her collector husband.

In the 1890s, Price organized several trips, usually consisting of students from Stanford University, to isolated areas in California, Arizona, and Baja California. On a trip to the Huachuca Mountains in February 1893, Ray L. Wilbur, a student recruited by Price, wrote of their adventure (W. K. Fisher 1923). Arriving in Tucson, they met "the editor [undoubtedly Herbert Brown] of one of the papers, who was interested in birds." On his advice, they went to nearby Fort Lowell before moving on to the southeastern mountains. "The country at that time was full of interesting characters. 'Apache Kid' [ca. 1869–?] had been loose in the Chiricahuas and we were none too comfortable at times in that portion of the Huachucas that extended over into Mexico." Nevertheless, Price proved to be a "delightful companion" and his "enthusiasm was contagious," which made for a most pleasurable trip. During the same period, another Stanford student, Wilfred H. Osgood (1903), compiled a list of birds in nearby Sulfur Springs Valley.

It was in 1893 that Loye "Padre" Miller (1874–1970), who knew "Billy" Price as a "grammar-school kid," joined him in southeastern Arizona Territory. Leaving Riverside, California, he traveled in a "rattletrap 'S.P.' car" to Tucson where, at 4:00 A.M. in "pitch dark," he met Billy, a hardly recognizable "husky chap in overalls and shoebrush whiskers."

The two set out immediately for Fort Lowell, where Miller (1950) remained for ten weeks. He later wrote of his first early morning impression in this

> *exciting new world, the desert of Arizona. . . . Three inches of snow*
> *had fallen the day before, and the air was snappy and perfectly clear.*
> *As dawn slowly came on, a thousand Palmer [Curve-billed] thrashers*
> *awoke into song, light began to kindle on the snow-powdered cliffs of*
> *the Catalina Mountains, and the Mexicans along the Rillito began to*
> *rouse, sending up the smoke of their cook fires built of desert mesquite*
> *wood. The rain-freshened desert, the smell of the creosote bush and*

mesquite smoke, the tingle of early morning air, and the spring song
of those . . . thrashers rounded out a picture that is still fresh in my
memory after fifty-odd years. I did not know it, but I was beginning
my professional life as a field naturalist.

Miller, as a young man open to the natural sights and sounds of his new surroundings, found his field escapades extremely thrilling along Rillito Creek. "The first song of the Scott oriole brought me up short. It sounded like a delirious meadowlark singing from the wooded bottom of the Rillito. And the first white-winged dove that turned loose his incredible song upon his arrival from Mexico!—it sounded like an inebriate owl on a jamboree . . . oh, so many new birds. So many new plants, too . . . and above all, the gigantic saguaro. I never got used to those great saguaros, nor have I to this day."

Later that summer, Price transferred Miller to the Chiricahua Mountains where again, he was "much impressed by the enormous numbers of band-tailed pigeons here. They came hurtling down the canyon at certain times of day to feed on the sweet acorns *(Quercus emoreyi),* which again was a new thing for me."

The following year in 1894, Price (1895) returned to the Chiricahuas, where he found his first nest of the Olive Warbler on "a small horizontal limb partially concealed by pine needles." He also explored the Graham and White Mountains.

Exploring wild places always held a fascination for Price. In 1896, he was again joined by Miller on a trip to the Cape Region of Baja California. Four years later, Price (1899) came close to losing his life in an adventure that took him down the Colorado River through the delta region to the Gulf of California. It was an arduous experience, for he lost his way and had no food and very little water before stumbling into a friendly Indian settlement, where he received food and help.

In later life, Miller, as a professor, "was able to instill in his students the same desire to search and to learn that he himself felt so keenly" (Howard 1971). He was chairman of the biology department at the University of California, Los Angeles, and will be remembered especially for his work on avian paleontology and oceanic birds of the west coast, the Gulf, and Baja California.

An interpretive naturalist, Miller revealed with great wit his virtuosity as an imitator of bird calls and songs on a most unusual recording, "Music in Nature," produced by the Cooper Ornithological Society in 1952. Among the birds he most enjoyed imitating and held a great fondness for was the Great Horned Owl, which he often mimicked on his student field trips.

In 1896, Francis C. Willard (1874–1930), a naturalist from Illinois, settled in Tombstone, where he taught school and helped an uncle operate a general supply store. For more than twenty years, he photographed birds and collected eggs, principally in the Huachuca Mountains, the San Pedro River near Fairbanks, and Tucson.

Arthur Cleveland Bent (1864–1954), a man who had considerable involvement with ornithologists all over the country while complying the monumental *Life Histories of North American Birds* series (1919–1958), spent two months with Willard in southeastern Arizona. Oliver L. Austin, Jr. (Bent et al. 1968), successor to Bent, described him as suffering from "a permanent tremble in his right hand" that occurred after he fell "from a tree during a youthful egging exploit." From this incident, Bent (1930) fully understood the dangers and worth of a climbing field ornithologist when he wrote a tribute to Willard's many field abilities:

> *Endowed with a splendid physique, prodigious and remarkable agility, skill and resourcefulness, he was, without exception the most efficient man in the field of all the many with whom I have had collecting experience. He was the best climber I have ever seen, absolutely fearless, full of tireless energy and so well equipped with experience and ingenuity, that nothing ever daunted him. . . . His knowledge of Arizona birds was so thorough, that he seemed to know just where each pair of birds would nest. . . . His powers of observation were well trained; he knew the birds and their habitats thoroughly, could recognize all their call notes and was familiar with all their habits.*

In 1897, Willard (1923) wrote of a hazardous encounter climbing to a Buff-breasted Flycatcher nest in the Huachuca Mountains. "This was my first year in the West, and nearly every day was bringing new acquaintances . . . astride the first branch forty feet up . . . clasped my arms around the trunk of the tree preparatory to climbing higher. Something soft gave under my hand and I knew without looking that it was the nest."

Willard described the nest-building activities of this flycatcher, the smallest of the lovely empids. "The female works persistently and rapidly, but the nest requires a lot of material and she often takes ten days to build it. She sits on the

Francis C. Willard (1874–1930) examining a Buff-breasted Flycatcher's nest in the Huachuca Mountains in Arizona in 1907. (From Condor, *1923)*

nest for short periods before the eggs are laid, and also as the eggs are laid, but does not seem to make a real business of it until the set is complete."

Another group of young naturalists from southern California were taking note of these isolated borderland mountains. Among them was Harry S. Swarth (1878–1935). His mentor was English immigrant George F. Morcom (1845–1932), about whom Swarth wrote: "[his] greatest contribution toward ornithology lies in his aid and encouragement to others in the accomplishment of what they might not other wise have done" (Swarth 1934). Living in the Los Angeles house of Swarth's father and uncle, Morcom used his upstairs quarters to house his bird collection and natural history library. Swarth, while in grammar school, "had the run of rooms where bird skins and birds' eggs were being handled" and, in the surrounding countryside, they "learned California birds, one by one."

At least two southwestern field naturalists provided bird notes and observations for Morcom: Joseph L. Hancock (1887) in the Corpus Christi area, and Frank Stephens in southern California (Morcom 1887). In later years, Morcom, like many collector naturalists, found that "taking life for any purpose became increasingly distasteful" (Swarth 1934). His large collection was finally donated to the California Academy of Sciences in 1929.

In February 1896, acting on the advice of Major Charles Bendire, Swarth, at the age of eighteen, made his first trip to the Huachuca Mountains with O. W. Howard, W. B. Judson, and H. G. Rising in a two-horse farm wagon (Linsdale 1936; Mailliard 1937). In all, Swarth made six trips to Arizona Territory. In recalling the abundance of Gambel's Quail on his first memorable Arizona trip, Swarth (1935) wrote they "were rarely, if ever, out of sight or hearing. All day long and day after day they scurried across the road; the birds were there in such countless numbers and humanity was so nearly absent that any suggestion of a future scarcity of quail could hardly have been entertained." In 1930, on a trip by auto over "the same route, perhaps twenty quail were seen! Making all allowances for different modes of transportation, the two sets of observations show plainly enough how rapidly the Gambel Quail is following the path taken by so many other American game birds."

The "Birds of the Huachuca Mountains, Arizona" (1904), compiled by Swarth, is an annotated list of 194 bird species that he and his associates found in this border area. Commenting on the Greater Pewee, Swarth remarked "where if not seen, it can at least be heard almost everywhere." Taking special note of the vocalizations sung by this flycatcher, Swarth added that the origin of its local name was "derived from its cry, Jose Maria (pronounced, Ho-say Maria)." No doubt, this was the source of this description for its song, which appears in several current guide books.

Other major contributions by Swarth to the natural history of the region were: "A Distributional List of the Birds of Arizona" (1914) and "The Faunal Areas of Southern Arizona: A Study in Animal Distribution" (1929). An industrious scientist, Swarth was affiliated with the Field Columbian Museum in Chicago; the Museum of Vertebrate Zoology at Berkeley; the Museum of History, Science and Art in Los Angeles; and the California Academy of Sciences in San Francisco.

Howard (1900, 1902) remained in the Huachucas four summer seasons and, while working on his mining claims, he collected primarily bird eggs. Among those collected were Wild Turkey and Prairie Falcon. About the same time, several other egg collectors came to the mountains. O. C. Poling and Richard Lusk secured the first egg sets of the Sulphur-bellied Flycatcher north of the border (Bendire 1895). But it was Lusk and Howard who took special note of the items taken from the Arizona Walnut that were used by the flycatchers in their nest construction. Using cavities in Arizona Sycamore, the birds built "nests which are marvels of uniformity and simplicity as to materials, are made of the naturally-curved, dried leaf stems of the walnut, without a shred of lining of any kind" (Lusk 1899). Howard (1899) added "the finer stems were placed on the inside of the nest" and, with a photograph, he illustrated their arrangement.

The unraveling of the specific distinctions between two small owls in this border region was long in coming. The more common and widespread Western Screech-Owl *(Otus kennicottii)*, generally occurring at lower elevations, had been named for Major Robert Kennicott (1835–1866), Alaskan explorer and founder of the Chicago Academy of Science, by Daniel G. Elliot in 1867. The other was the very similar Whiskered Screech-Owl, a bird restricted to the pine-oak woodlands, and whose northern regional limits are in the nearby Santa Catalina Mountains. Although W. E. D. Scott (1886a) collected "Mexican Screech Owls" around Tucson and Santa Catalina Mountains up to forty-five hundred feet, he may not have secured both species, or if he did, he overlooked their subtle plumage differences and failed to recognize their easily identifiable vocal distinctions. Understandably, the northern range of this Mexican owl was not realized until William Brewster (1898) received two specimens taken from the Huachucas by Lusk. Although both species had been previously described, Brewster confirmed the presence of both owls and that the Whiskered Screech-Owl was "quite new to our fauna" north of the border.

During 1902 and 1903, Harry Swarth, accompanied by O. W. Howard and F. Stephens, explored the Santa Rita Mountains and the Santa Cruz River near the old San Xavier Mission ten miles south of Tucson. Their investigations along the Santa Cruz River were in an area of unspoiled desert diversity. Swarth (1905) described the scene: "The river, running underground for most of its course, rises to the surface at this point, and the bottom lands on either side are covered, miles in extent, with a thick growth of giant mesquite trees, literally giants . . . many of them sixty feet high and over. . . . This magnificent grove is included in the Papago Indian [San Xavier] reservation, which is the only reason for the trees surviving as long as they have, since elsewhere every mesquite large enough to be used as firewood has been ruthlessly cut down." Among the many rich and diverse desert streams that have since been dramatically impacted by a host of circumstances are the Gila (Rea 1983) and Colorado (Rosenberg et al. 1991) Rivers.

Continuing, Swarth added that the "medley of bird songs was absolutely confusing" because of the abundance of individuals and species supported within this luxuriant forest. But, alas, as in the case of many southwestern streams, Herbert Brandt (1951), when visiting the "grand mesquite forest" three decades later, "could hear the thump, thump, of . . . man's ruinous ax." Brandt returned in 1945 and, with "sad heart . . . found that every noble tree had been hacked away for firewood."

Epilogue

"It is to be hoped that it [Passenger Pigeon] may, in time, have some little
influence in staying the hand of man in his career of extermination of
such incomparably beautiful creatures in nature. I, for one, greatly fear
that this will not be the case. In due course, the day will come when
practically all of the world's avifauna will have become utterly extinct.
Such a fate for it is coming to pass now, with far greater rapidity than
most people realize."
—ROBERT W. SHUFELDT (1914)

DURING THE PERIOD addressed in this book, ornithology was a science
in need of handheld subjects or specimens to study and with which to further
our knowledge. To this end, the collecting of birds by the early naturalists was
the first step in obtaining information about these living subjects about which
so little was known. The justification for taking specimens by the early naturalists was most assertively put into proper context by Vernon Bailey. "Our present
knowledge of birds and their classifications has come from a study of specimens
... without this foundation the study of birds would not have its deep interest
and meaning nor its practical bearing on the economy of our lives. Even our
enjoyment of the birds in life, their beauty, song, and friendship, would be far
less than it is without the underlying knowledge of their life history, the place
they fill, and their importance to us" (F. M. Bailey 1902).

To the early naturalist the only credible record was a "bird in the hand."
However, with improved camera and binocular optics, reasonable certainty of

species identification was eventually recognized. In most circumstances these effective tools have replaced the collector's gun. Roger Tory Peterson (1965), in crediting Ludlow Griscom (1890–1965) as the first observer who "bridged the gap" between early collectors and modern field ornithologists, remarked: "[Griscom] demonstrated that practically all birds have their field marks and that it is seldom necessary for a trained man to shoot a bird to know, at the specific level, precisely what it is."

Today, with the additional techniques of bird netting and banding, together with protective laws, the practice of securing specimens has largely been eliminated. Most collections have also been absorbed from private to public institutions, such as universities, museums, state game agencies, and the U.S. Fish and Wildlife Service. Utilized not only for their scientific value and educational exhibits, specimens have also served as important bench marks for comparing shell thickness and species deformation resulting from the accumulative effects of pesticides and other toxins.

Birds, as we have noted, were first brought to our attention in a serious way by the extraordinary curiosity of early naturalists. But, with the close of the nineteenth century, ornithology began a complex metamorphosis. Field investigation, the cornerstone of ornithology, continued, but with it came a wide range of ideas promoting extensive research. From ornithology have emerged avian components of study encompassing paleontology, physiology, systematic and molecular phylogeny, behavior, song, migration, and several others (Streseman 1975). Many of these branches, while in their infancy, were briefly touched on by the early naturalists. For example, from Darwin's ([1859] 1964) remarks "that species in a state of nature are limited in their ranges by the competition of other organic beings" has emerged the study of ecology.

As the twenty-first century begins, the world phylogenetic "tree" is still in ferment, and it is doubtful that it will ever achieve maturity. Continuing taxonomic and nomenclature changes resulting from DNA-DNA hybridization studies have been adopted by the 1997 AOU Committee on Classification and Nomenclature with the realization that "we have barely begun the reorganization of our concepts of relationships among avian groups." This committee, recognized as the scientific governing body on this subject, has commendably carried forth the maintenance of avian systematics from their early predecessors with minimal modifications. Its findings form the basis that serve as the standard reference. Still, understanding and interpreting the speciation concept with all its complexities remains perhaps as Darwin ([1859] 1964) stated: "No one definition has as yet satisfied all naturalists; yet every naturalist knows vaguely what he means when he speaks of a species."

By the end of the nineteenth century, most bird types occurring in the Southwest had been collected, but their taxon placement and their naming in some instances is still subject to revision. Continual revision within the naming process has on occasion effected the loss of eponyms. While some prominent naturalists have visibly disappeared from the AOU Check-list, others have surfaced. As noted in the text, the traditional considerations prompting nomenclature changes have been largely functions of priority or choice due to taxonomic guidelines. By understanding the precepts of this process along with the retention of eponyms where possible, there will, no doubt, be a greater appreciation of ornithological history.

Naturalists will always delve into former times in an effort to visualize and understand natural systems and conditions that have preceded us. In writing about the 1832–1834 western North American visit of Prussian naturalist Alexander Philipp Maximilian, Prince of Wied-Nuewied, Vernon Bailey (1923) expressed concerns, seventy-five years ago, that still apply to the present. "As our wild life and primitive conditions disappear, such records of the New World before the newness was gone have an ever increasing value and should be more generally known."

In scanning the twenty-first century horizon with some trepidation, let us trust that the avaricious hand of humans, as ruefully predicted by Shufeldt's epilogue quote, will never be realized as a result of an insouciant society. Our natural inheritance is our responsibility to safeguard. Certainly the planet is diminished every time there is a significant reduction or extinction of a living component. But even as some elements of our complex natural world decrease, perhaps the lives of early naturalists, partially owing to shrinking biological vistas, will enjoy a greater, although sobering, review.

True to the inquisitive spirit of many early naturalists, the unpublished reflection of Professor Loye "Padre" Miller, inscribed by him in his *Lifelong Boyhood* (1950) to a student, is fitting for the conclusion of this book: "The world is so full of such interesting things and it *is* such fun finding out about them! Don't ever lose sight of that enjoyment. If you do you've grown old."

*A*PPENDIX

NEXUS OF FIRST OBSERVERS, COLLECTORS, DESCRIBERS, AND EPONYMS OF BIRDS RELEVANT TO THE SOUTHWEST.

The criteria determining which birds are included in the appendix relate primarily to their discovery in the borderland or adjacent regions. Other birds noted in this area but discovered elsewhere are omitted. First observations or paintings by Spanish conquistadores, explorers, and missionaries that imply reasonable species certainty are also noted. A bird sighting without securing a specimen was not sufficient to authenticate a discovery. Collectors generally recognized new species in the field or sent them to museums to be examined. Describers or systematists then announced new discoveries in publications. Not all collectors were noted in the literature. Eponyms applied to the earliest descriptions have generally been retained by the rule of priority. Listings also include many subspecies eponyms (Banks 1987). If the describer's name is in parentheses, the species was originally described in another genus.

Individuals involved in the discovery and naming of birds included many nationalities. They were Austrian [A], Belgian [B], Dalmatian [Da], Dutch [D], English [E], French [F], German [G], Hungarian [H], Irish [Ir], Italian [I], Mexican [M], Norwegian [N], Scottish [Sc], Spanish [Sp] and Swedish [S]. Russia, like many nations, sponsored naturalists from other countries. + denotes became U.S. citizen. * denotes current eponym. No letter denotes birth as U.S. citizen.

COMMON AND SCIENTIFIC NAMES	FIRST OBSERVER	COLLECTOR	EARLIER EPONYM	DESCRIBER/DATE
Albatross, Short-tailed *Phoebastria albatrus*	—	—	Steller's	(Pallas) [G] 1769
Ani, Grooved-billed *Crotophaga sulcirostris*	Sessé [Sp]/ Moziño [M]	Bullock [E]	—	Swainson [E] 1827
Auklet, Cassin's* *Ptychoramphus aleuticus*	—	—	—	(Pallas) [G] 1811
Beardless-Tyrannulet, Northern *Camptostoma imberbe*	—	Sallé [F]	—	Sclater [E] 1857
Becard, Rose-throated *Pachyramphus aglaiae**	—	—	Xantus'	(Lafresnaye) [F] 1839
Blackbird, Brewer's* *Euphagus cyanocephalus*	—	Deppe [G] ?	—	(Wagler) [G] 1829
Blackbird, Tricolored *Agelaius tricolor*	—	Nuttall [E]	—	(Audubon, J.J.) 1837
Blackbird, Yellow-headed *Xanthocephalus xanthocephalus*	Sessé [Sp]/ Moziño [M]	Say/Peale	—	(Bonaparte) [I] 1826
Black-Hawk, Common *Buteogallus anthracinus*	Sessé [Sp]/ Moziño [M]	Deppe [G]	—	(Bonaparte) [I] 1826
Bluebird, Western *Sialia mexicana*	—	—	—	Swainson [E] 1832

Species	Author	Eponym	Collector 1	Collector 2
Bobwhite, Northern *Colinus virginianus*	(Linnaeus) [S] 1758	Grayson's; *graysoni*	Catesby [E]	Vaca [Sp]
Booby, Blue-footed *Sula nebouxii**	Milne-Edwards [F] 1882	—	Néboux [F]	—
Booby, Brown *Sula leucogaster*	(Boddaert) [?] 1783	Brewster's; *brewsteri*	—	—
Bunting, Blue *Cyanocompsa parellina*	(Bonaparte) [I] 1850	—	Deppe [G]	Sessé [Sp]/ Moziño [M]
Bunting, Lark *Calamospiza melanocorys*	Stejneger [N]+ 1885	—	Townsend, J. K.	—
Bunting, Lazuli *Passerina amoena*	(Say) 1823	—	Say/Peale	—
Bunting, Varied *Passerina versicolor*	(Bonaparte) [I] 1838	—	Paris brothers [F or I]	—
Bushtit *Psaltriparus minimus*	(Townsend, J. K.) 1837	—	Townsend, J. K.	—
Caracara, Crested *Caracara plancus*	(Miller) [E] 1777	Audubon's	—	—
Caracara, Guadalupe *Caracara lutosus* [1]	(Ridgway) 1876	—	Palmer [E]+	—
Chachalaca, Plain *Ortalis vetula*	(Wagler) [G] 1830	—	Deppe [G]	Pfefferkorn [G]

COMMON AND SCIENTIFIC NAMES	FIRST OBSERVER	COLLECTOR	EARLIER EPONYM	DESCRIBER/DATE
Chat, Tres Marias *Granatellus francescae**	—	Grayson	—	Baird 1865
Chickadee, Mexican *Poecile sclateri**	—	Sallé [F]	—	(Kleinschmidt) [G] 1897
Chickadee, Mountain *Poecile gambeli**	—	Gambel	Bailey's; Mrs. Bailey's; Barlow's; Grinnell's	(Ridgway) 1886
Condor, California *Gymnogyps californianus*	Ascension [Sp]	Menzies [Sc]	—	(Shaw) [E] 1798
Cormorant, Brandt's* *Phalacrocorax penicillatus*	—	—	—	(Brandt) [G] 1837
Cowbird, Bronzed *Molothrus aeneus*	—	Deppe [G] ?	—	(Wagler) [G] 1829
Cowbird, Brown-headed *Molothrus ater*	Pfefferkorn [G] ?	—	—	(Boddaert) [?] 1783
Crane, Sandhill *Grus canadensis*	Coronado [Sp]	Edwards [E]	—	(Linnaeus) [S] 1758
Crossbill, Red *Loxia curvirostra*	—	—	Bendire's; Bent's; Grinnell's	Linnaeus [S] 1758
Dipper, American *Cinclus mexicanus*	—	Bullock [E]	—	Swainson [E] 1827

Species				
Dove, Inca *Columbina inca*	—	—	—	(Lesson) [F] 1847
Dove, Mourning *Zenaida* macroura*	—	Edwards [E]	—	(Linnaeus) [S] 1758
Dove, Socorro *Zenaida* graysoni*²	—	Grayson	—	(Lawrence) 1871
Dove, White-tipped *Leptotila verreauxi**	—	—	—	Bonaparte [I] 1855
Dove, White-winged *Zenaida* asiatica*	—	Edwards [E]	—	(Linnaeus) [S] 1758
Dowitcher, Long-billed *Limnodromus scolopaceus*	—	Say/Peale	—	(Say) 1823
Eagle, Bald *Haliaeetus leucocephalus*	Coronado [Sp]	Catesby [E]	—	(Linnaeus) [S] 1766
Eagle, Golden *Aquila chrysaetos*	Coronado [Sp]	—	—	(Linnaeus) [S] 1758
Egret, Snowy *Egretta thula*	—	—	Brewster's	(Molina) [I] 1782
Falcon, Aplomado *Falco femoralis*	Sessé [Sp]/ Moziño [M]	—	—	Temminck [D] 1822
Falcon, Peregrine *Falco peregrinus*	—	—	Peale's	Tunstall [E] 1771

COMMON AND SCIENTIFIC NAMES	FIRST OBSERVER	COLLECTOR	EARLIER EPONYM	DESCRIBER/DATE
Falcon, Prairie *Falco mexicanus*	Sessé [Sp]/Moziño [M]	—	—	Schlegel [G] 1850
Finch, Cassin's* *Carpodacus cassinii**	—	Kennerly/ Möllhausen [G]	Cassin's Purple	Baird 1854
Finch, House *Carpodacus mexicanus*	—	—	McGregor's; *mcgregori*	(Müller) [G] 1776
Flicker, Gilded *Colaptes chrysoides*	—	—	Malherbe's; Mearn's Gilded	(Malherbe) [F] 1852
Flycatcher, Ash-throated *Myiarchus cinerascens*	—	—	McCown	(Lawrence) 1851
Flycatcher, Brown-crested *Myiarchus tyrannulus*	—	—	Wied's Crested	(Müller) [G] 1776
Flycatcher, Buff-breasted *Empidonax fulvifrons*	—	—	—	(Giraud) 1841
Flycatcher, Cordilleran *Empidonax occidentalis*	—	Nelson/Goldman	—	Nelson 1897
Flycatcher, Dusky *Empidonax oberholseri**	—	Phillips	Wright's	Phillips 1939
Flycatcher, Gray *Empidonax wrightii**	—	Wright	Wright's; *oberholseri*	Baird 1858

Species				
Flycatcher, Hammond's* *Empidonax hammondii**	—	Xántus [H]+	—	(Xántus) [H]+ 1858
Flycatcher, Nutting's* *Myiarchus nuttingi**	—	Nutting	—	Ridgway 1883
Flycatcher, Olive-sided *Contopus cooperi**	—	—	—	(Nuttall) [E]1831
Flycatcher, Pacific-slope *Empidonax difficilis*	—	Cooper/ Xántus [H]+	—	Baird 1858
Flycatcher, Scissor-tailed *Tyrannus forficatus*	—	—	—	(Gmelin) [G] 1789
Flycatcher, Sulphur-bellied *Myiodynastes luteiventris*	—	Sallé [F] ?	—	Sclater [E] 1859
Flycatcher, Vermilion *Pyrocephalus rubinus*	—	—	—	(Boddaert) [?] 1783
Flycatcher, Willow *Empidonax traillii**	—	Audubon, J.J.	Traill's	(Audubon, J.J.) 1828
Gnatcatcher, Black-capped *Polioptila nigriceps*	—	Xántus [H]+	—	Baird 1864
Gnatcatcher, Black-tailed *Polioptila melanura*	—	McCown?	—	Lawrence 1857
Gnatcatcher, California *Polioptila californica*	—	Stephens	Xantus'	Brewster 1881

COMMON AND SCIENTIFIC NAMES	FIRST OBSERVER	COLLECTOR	EARLIER EPONYM	DESCRIBER/DATE
Goldeneye, Barrow's* / Bucephala islandica	—	—	—	(Gmelin) [G] 1789
Goldfinch, Lawrence's* / Carduelis lawrencei*	—	Bell	—	Cassin 1850
Goldfinch, Lesser / Carduelis psaltria	Sessé [Sp]/Moziño [M]	Say/Peale	—	(Say) 1823
Goose, Canada / Branta canadensis	Catesby [E]	—	—	(Linnaeus) [S] 1758
Goose, Ross's* / Chen rossii*	Hearne [E]	—	—	(Cassin) 1861
Goose, Snow / Chen caerulescens	—	—	—	(Linnaeus) [S] 1758
Grackle, Great-tailed / Quiscalus mexicanus	—	—	—	(Gmelin) [G] 1788
Grebe, Clark's* / Aechmophorus clarkii*	—	Clark, J. H.	—	(Lawrence) 1858
Grebe, Western / Aechmophorus occidentalis	—	Trowbridge	—	(Lawrence) 1858
Grosbeak, Black-headed / Pheucticus melanocephalus	—	Bullock [E]	—	(Swainson) [E] 1827

Species				
Grouse, Blue *Dendragapus obscurus*	—	Say/Peale	—	(Say) 1823
Gull, Bonaparte's* *Larus philadelphia*	—	—	—	(Ord) 1815
Gull, California *Larus californicus*	—	Holden	—	Lawrence 1854
Gull, Franklin's* *Larus pipixcan*	—	Deppe [G] ?	—	Wagler [G] 1831
Gull, Heermann's* *Larus heermanni**	—	Heermann	—	Cassin 1852
Gull, Thayer's* *Larus thayeri**	—	—	—	Brooks 1915
Gull, Western *Larus occidentalis*	—	Townsend, J. K.	—	Audubon, J. J. 1839
Gull, Yellow-footed *Larus livens*	—	Dwight	—	Dwight 1919
Hawk, Cooper's* *Accipiter cooperii**	Sessé [Sp]/Moziño [M] ?	—	—	(Bonaparte) [I] 1828
Hawk, Ferruginous *Buteo regalis*	—	Deppe [G]	—	(Gray) [E] 1844
Hawk, Harris's* *Parabuteo unicinctus*	—	—	Harris's Buzzard	(Temminck) [D] 1824

COMMON AND SCIENTIFIC NAMES	FIRST OBSERVER	COLLECTOR	EARLIER EPONYM	DESCRIBER/DATE
Hawk, Red-tailed *Buteo jamaicensis*	—	—	Cooper's Buzzard; Harlan's Buzzard; Krider's Buzzard	(Gmelin) [G] 1788
Hawk, Swainson's* *Buteo swainsoni**	—	Townsend, J. K.	Swainson's Buzzard	Bonaparte [I] 1838
Hawk, White-tailed *Buteo albicaudatus*	Sessé [Sp]/Moziño [M]	—	Sennett's White-tailed	Vieillot [F] 1816
Hawk, Zone-tailed *Buteo albonotatus*	—	—	—	Kaup [G?] 1847
Heron, Great Blue *Ardea herodias*	—	Edwards [E]	Ward's	Linnaeus [S] 1758
Heron, Green *Butorides virescens*	—	Catesby [E]	Anthony's Green; Frazar's Green	(Linnaeus) [S] 1758
Hummingbird, Allen's* *Selasphorus sasin*	—	—	—	(Lesson) [F] 1829
Hummingbird, Anna's* *Calypte anna**	—	Botta [I]	—	(Lesson) [F] 1829
Hummingbird, Berylline *Amazilia beryllina*	—	—	—	(Deppe) [G] 1830
Hummingbird, Black-chinned *Archilochus alexandri**	—	Alexandre [M?]	Alexander	(Bourcier [F]/ Mulsant) [F] 1846

Species				
Hummingbird, Blue-throated *Lampornis clemenciae*	Sessé [Sp]/Moziño [M]	Botta [I]	—	(Lesson) [F] 1829
Hummingbird, Broad-billed *Cynanthus latirostris*	—	Bullock [E]	—	Swainson [E] 1827
Hummingbird, Broad-tailed *Selasphorus platycercus*	—	Bullock [E]	—	(Swainson) [E] 1827
Hummingbird, Buff-bellied *Amazilia yucatanensis*	—	—	—	(Cabot) 1845
Hummingbird, Bumblebee *Atthis heloisa**	—	—	Heloise's; Morcom's	(DeLattre [F]/ Lesson) [F] 1839
Hummingbird, Calliope *Stellula calliope*	—	Floresi [I]	—	(Gould) [E] 1847
Hummingbird, Costa's* *Calypte costae**	—	Néboux [F]	—	(Bourcier) [F] 1839
Hummingbird, Lucifer *Calothorax lucifer*	—	Bullock [E]	—	(Swainson) [E] 1827
Hummingbird, Magnificent *Eugenes fulgens*	—	Bullock [E]	Rivoli's	(Swainson) [E] 1827
Hummingbird, Rufous *Selasphorus rufus*	—	Cook [E]/ Anderson [E]	—	(Gmelin) [G] 1788
Hummingbird, Rufous-tailed *Amazilia tzacatl*	—	—	Rieffer's	(De la Llave) [M] 1833

COMMON AND SCIENTIFIC NAMES	FIRST OBSERVER	COLLECTOR	EARLIER EPONYM	DESCRIBER/DATE
Hummingbird, Violet-crowned *Amazilia violiceps*	—	Deppe [G] ?	—	(Gould) [E] 1859
Hummingbird, White-eared *Hylocharis leucotis*	—	—	—	(Vieillot) [F] 1818
Hummingbird, Xantus's* *Hylocharis xantusii**	—	Xántus [H]+	—	(Lawrence) 1860
Jabiru *Jabiru mycteria*	Sessé [Sp]/Moziño [M]	—	—	(Lichtenstein) [G] 1819
Jay, Brown *Cyanocorax morio*	—	Deppe [G] ?	—	(Wagler) [G] 1829
Jay, Green *Cyanocorax yncas*	—	—	—	(Boddaert) [?] 1783
Jay, Mexican *Aphelocoma ultramarina*	Sessé [Sp]/Moziño [M]	—	Couch's; Sieber's	(Bonaparte) [I] 1825
Jay, Pinyon *Gymnorhinus cyanocephalus*	Lewis/ Clark, W.	Wied [G]	Maximillian's	Weid [G] 1841
Jay, Steller's* *Cyanocitta stelleri**	—	Steller [G]	Grinnell's; Osgood's	(Gmelin) [G] 1788
Junco, Dark-eyed *Junco hyemalis*	—	Catesby [E]	Baird's; Ridgway's; Shufeldt's; Thurber's; Townsend's	(Linnaeus) [S] 1758

	Sessé [Sp]/Moziño [M]	Deppe [G]		
Junco, Yellow-eyed *Junco phaeonotus*	—	?	—	Wagler [G] 1831
Kingbird, Cassin's* *Tyrannus vociferans*	—	—	—	Swainson [E] 1826
Kingbird, Couch's* *Tyrannus couchii**	—	Couch	—	Baird 1858
Kingbird, Thick-billed *Tyrannus crassirostris*	—	—	—	Swainson [E] 1826
Kingbird, Tropical *Tyrannus melancholicus*	—	—	Couch's; Lichtenstein's	Vieillot [F] 1819
Kingbird, Western *Tyrannus verticalis*	—	Say/Peale	—	Say 1823
Kingfisher, Belted *Ceryle alcyon*	—	Catesby [E]	—	(Linnaeus) [S] 1758
Kingfisher, Green *Chloroceryle americana*	—	—	Cabanis'	(Gmelin) [G] 1788
Kingfisher, Ringed *Ceryle torquata*	—	—	—	(Linnaeus) [S] 1766
Kinglet, Golden-crowned *Regulus satrapa*	—	—	—	Lichtenstein [G] 1823
Kiskadee, Great *Pitangus sulphuratus*	—	—	Lord Derby's	(Linnaeus) [S] 1766

COMMON AND SCIENTIFIC NAMES	FIRST OBSERVER	COLLECTOR	EARLIER EPONYM	DESCRIBER/DATE
Longspur, Chestnut-collared *Calcarius ornatus*	—	Townsend, J. K.	—	(Townsend, J. K.) 1837
Longspur, McCown's* *Calcarius mccownii**	—	McCown	McCown's Lark Bunting	(Lawrence) 1851
Longspur, Smith's* *Calcarius pictus*	—	—	—	(Swainson) [E] 1832
Loon, Pacific *Gavia pacifica*	—	Suckley/ Trowbridge	—	(Lawrence) 1858
Macaw, Scarlet *Ara macao*	Espejo [Sp]	—	—	(Linnaeus) [E] 1758
Magpie, Yellow-billed *Pica nuttalli**	—	Nuttall [E]	Nuttall's	(Audubon, J.J.) 1837
Mallard *Anas platyrhynchos*	—	—	*diazi*	Linnaeus [S] 1758
Meadowlark, Western *Sturnella neglecta*	Lewis/ Clark	Audubon, J.J.	—	Audubon, J. J. 1844
Merlin *Falco columbarius*	—	Catesby [E]	Richardson's Pigeon Hawk	Linnaeus [S] 1758
Mockingbird, Blue *Melanotis caerulescens*	—	Bullock [E]	—	(Swainson) [E] 1827

Species				
Mockingbird, Northern *Mimus polyglottos*	—	Catesby [E]	—	(Linnaeus) [S] 1758
Mockingbird, Socorro *Mimodes graysoni**	—	Grayson	—	(Lawrence) 1871
Murrelet, Craveri's* *Synthliboramphus craveri**	—	Craveri [I]	—	(Salvadori) [I] 1865
Murrelet, Xantus's* *Synthliboramphus hypoleucus*	—	Xántus [H]+	—	(Xántus) [H]+1860
Nighthawk, Common *Chordeiles minor*	—	Catesby [E]	Asseri; Cherrie's; Sennett's	(Forster) [G] 1771
Nighthawk, Lesser *Chordeiles acutipennis*	—	—	—	(Hermann) [A] 1783
Nightjar, Buff-collared *Caprimulgus ridgwayi**	Sessé [Sp]/Moziño [M]	Nelson/ Goldman	Ridgway's Whip-poor-will	(Nelson) 1897
Nutcracker, Clark's* *Nucifraga columbiana*	—	Lewis/ Clark	Clark's Crow	(Wilson) [Sc] 1811
Nuthatch, Pygmy *Sitta pygmaea*	—	Beechey [E]	—	Vigors [Ir] 1839
Oriole, Altamira *Icterus gularis*	—	Deppe [G]	Lichtenstein's	(Wagler) [G] 1829
Oriole, Audubon's* *Icterus graduacauda*	—	—	—	Lesson [F] 1839

COMMON AND SCIENTIFIC NAMES	FIRST OBSERVER	COLLECTOR	EARLIER EPONYM	DESCRIBER/DATE
Oriole, Black-backed* *Icterus abeillei**	—	—	Abeillé's	(Lesson) [F] 1839
Oriole, Black-vented* *Icterus wagleri**	—	Sallé [F]	Wagler's	Sclater [E] 1857
Oriole, Bullock's* *Icterus bullockii**	—	Bullock [E]	—	(Swainson) [E] 1827
Oriole, Hooded *Icterus cucullatus*	—	Bullock [E]	Nelson's; Sennett's	Swainson [E] 1827
Oriole, Orchard *Icterus spurius*	—	Catesby [E]	Fuertes'	(Linnaeus) [S] 1766
Oriole, Scott's* *Icterus parisorum**	—	Paris brothers [F or I]	—	Bonaparte [I] 1838
Oriole, Streaked-backed *Icterus pustulatus*	—	Deppe [G] ?	—	(Wagler) [G] 1829
Owl, Elf *Micrathene whitneyi**	—	Cooper	Whitney's Elf	(Cooper) 1861
Owl, Flammulated *Otus flammeolus*	—	—	—	(Kaup) [G?] 1853
Owl, Northern Saw-whet *Aegolius acadicus*	—	—	Kirtland's	(Gmelin) [G] 1788

Common name / Scientific name			Alternate name	Authority, date
Owl, Spotted / *Strix occidentalis*	—	Xántus [H]+	—	(Xántus) [H]+ 1860
Oystercatcher, American / *Haematopus palliatus*	—	—	Frazar's	Temminck [D] 1820
Oystercatcher, Black / *Haematopus bachmani**	—	Townsend, J. K.	—	Audubon, J. J. 1838
Parrot, Thick-billed / *Rhynchopsitta pachyrhyncha*	Espejo [Sp]	Bullock [E]	—	(Swainson) [E] 1827
Parula, Tropical / *Parula pitiayumi*	—	—	Sennett's Warbler	(Vieillot) [F] 1817
Pauraque, Common / *Nyctidromus albicollis*	—	—	Merrill's	(Gmelin) [G] 1789
Pewee, Greater / *Contopus pertinax*	—	Deppe [G] ?	Coues' Flycatcher; Coues'	Cabanis [G]/ Heine [G] 1859
Phainopepla / *Phainopepla nitens*	—	—	—	(Swainson) [E] 1838
Phalarope, Wilson's* / *Phalaropus tricolor*	—	—	—	(Vieillot) [F] 1819
Phoebe, Black / *Sayornis* nigricans*	Pfefferkorn [G] ?	Bullock [E]	—	(Swainson) [E]1827
Phoebe, Eastern / *Sayornis* phoebe*	—	—	—	(Latham) [E] 1790

COMMON AND SCIENTIFIC NAMES	FIRST OBSERVER	COLLECTOR	EARLIER EPONYM	DESCRIBER/DATE
Phoebe, Say's* / Sayornis* saya*	—	Say/Peale	Say's Flycatcher	(Bonaparte) [I] 1825
Pigeon, Band-tailed / Columba fasciata	Sessé [Sp]/ Moziño [M]	Say/Peale	Viosca's	Say 1823
Pigeon, Red-billed / Columba flavirostris	—	Deppe [G]	—	Wagler [G] 1831
Pipit, Sprague's* / Anthus spragueii*	—	Audubon, J.J.	Sprague's Lark	(Audubon, J.J.) 1844
Plover, Mountain / Charadrius montanus	—	Townsend, J.K.	—	Townsend, J.K. 1837
Plover, Wilson's* / Charadrius wilsonia*	Sessé [Sp]/Moziño [M]	—	Belding's	Ord 1814
Poorwill, Common / Phalaenoptilus nuttallii*	Lewis/ Clark	Audubon, J.J.	Nuttall's Whip-poor-will	(Audubon, J.J.) 1844
Prairie-Chicken, Greater / Tympanuchus cupido	—	Catesby[E]	Attwater's	(Linnaeus) [S] 1758
Prairie-Chicken, Lesser / Tympanuchus pallidicinctus	—	Pope/Diffenderfer	—	(Ridgway) 1873
Pygmy-Owl, Ferruginous / Glaucidium brasilianum	—	—	—	(Gmelin) [G] 1788

Species				
Pygmy-Owl, Northern *Glaucidium gnoma*	—	Petz [?] ?	Hoskins' Pygmy	Wagler [G] 1832
Pyrrhuloxia *Cardinalis sinuatus*	—	Paris brothers [F or I]	—	Bonaparte [I]1838
Quail, California *Callipepla californica*	Sessé [Sp]/Moziño [M]	Menzies [Sc]	Canfield's	(Shaw) [E] 1798
Quail, Elegant *Callipepla douglasii**	—	—	—	(Vigors) [Ir] 1829
Quail, Gambel's* *Callipepla gambelii**	Coronado [Sp]	Gambel	Gambel's Crested Partridge	(Gambel) 1843
Quail, Montezuma* *Cyrtonyx montezumae**	—	Petz [?] ?	Massena Partridge; Mearn's	(Vigors) [Ir] 1830
Quail, Mountain *Oreortyx pictus*	Lewis/ Clark	Douglas [Sc]	—	(Douglas) [Sc] 1829
Quail, Scaled *Callipepla squamata*	Vaca [Sp]	Petz [?] ?	—	(Vigors) [Ir] 1830
Rail, Clapper *Rallus longirostris*	—	—	Belding	Boddaert [?] 1783
Raven, Chihuahuan *Corvus cryptoleucus*	Pfefferkorn [G]	Couch	—	Couch 1854
Redstart, Painted *Myioborus pictus*	—	—	—	(Swainson) [E] 1829

COMMON AND SCIENTIFIC NAMES	FIRST OBSERVER	COLLECTOR	EARLIER EPONYM	DESCRIBER/DATE
Roadrunner, Greater *Geococcyx californianus*	La Pérouse [F]; Pfefferkorn [G]	Botta [I]	—	(Lesson) [F] 1829
Robin, Clay-colored *Turdus grayi**	—	Leon [?]	Gray's Thrush	Bonaparte [I] 1838
Robin, Grayson's* *Turdus graysoni**	—	Grayson	—	Ridgway 1882
Robin, Rufous-backed *Turdus rufopalliatus*	—	—	—	Lafresnaye [F] 1840
Robin, White-throated *Turdus assimilis*	—	Deppe [G] ?	—	Cabanis [G] 1850
Rosy-Finch, Black *Leucosticte atrata*	—	—	Hepburn's Leucosticte; Hepburn's Rosy	Ridgway 1874
Sandpiper, Baird's* *Calidris bairdii**	—	—	—	(Coues) 1861
Sandpiper, Western *Calidris mauri**	—	Cabanis [G]	—	(Cabanis) [G] 1857
Sapsucker, Williamson's* *Sphyrapicus thyroideus*	—	Bell/ Newberry, J. S.	Natalie's; Williamson's Woodpecker	(Cassin) 1852

Species				
Screech-Owl, Western *Otus kennicottii*	—	Bischoff [G]	McCall's Screech; Aiken's Screech; Bancroft's Screech; Brewster's Screech; Gilman's Screech; Grinnell's Screech; Hasbrouck's Screech; Kennicott's Screech; Xantus' Screech	(Elliot) 1867
Screech-Owl, Whiskered *Otus trichopsis*	—	Petz [?] ?	—	(Wagler) [G] 1832
Scrub-Jay, Island *Aphelocoma insularis*	—	Henshaw	—	Henshaw 1886
Scrub-Jay, Western *Aphelocoma californica*	Sessé [Sp]/Moziño [M]	Beechey [E]	Belding's; Woodhouse's; Xantus'	(Vigors) [Ir] 1839
Shearwater, Audubon's* *Puffinus lherminieri**	—	—	—	Lesson [F] 1839
Shearwater, Black-vented *Puffinus opisthomelas*	—	Xántus [H]+	—	Coues 1864
Shearwater, Pink-footed *Puffinus creatopus*	—	Cooper	—	Coues 1864
Shearwater, Townsend's* *Puffinus auricularis*	—	Townsend, C. H.	Newell's	Townsend, C. H. 1890

COMMON AND SCIENTIFIC NAMES	FIRST OBSERVER	COLLECTOR	EARLIER EPONYM	DESCRIBER/DATE
Shrike, Loggerhead *Lanius ludovicianus*	—	—	Gambel; Grinnell's; Nelson's	Linnaeus [S] 1766
Silky-flycatcher, Gray *Ptilogonys cinereus*	—	—	—	Swainson [E] 1827
Snipe, Common *Gallinago gallinago*	—	—	Wilson's	(Linnaeus) [S] 1758
Solitaire, Townsend's* *Myadestes townsendi*	—	Townsend, J. K.	Townsend's Flycatcher; Townsend's Ptilogonys	(Audubon, J. J.) 1838
Sparrow, Baird's* *Ammodramus bairdii*	—	Audubon, J. J.	Baird's Bunting	(Audubon, J. J.) 1844
Sparrow, Black-chinned *Spizella atrogularis*	Sessé [Sp]/Moziño [M]	—	—	(Cabanis) [G] 1851
Sparrow, Black-throated *Amphispiza bilineata*	—	Audubon, J. W.	Bangs'	(Cassin) 1850
Sparrow, Botteri's* *Aimophila botterii*	—	Botteri [Da]	—	(Sclater) [E] 1858
Sparrow, Brewer's* *Spizella breweri*	—	—	Brewer's Chipping	Cassin 1856
Sparrow, Cassin's* *Aimophila cassinii*	—	Woodhouse	Cassin's Summer Finch	(Woodhouse) 1852

Sparrow, Five-striped *Aimophila quinquestriata*	—	Salvin [E]	—	(Sclater [E]/ Salvin [E]) 1868
Sparrow, Fox *Passerella iliaca*	—	—	Stephens's Fox; Townsend's Fox	(Merrem [G]) 1786
Sparrow, Harris's* *Zonotrichia querula*	—	Nuttall [E]	—	(Nuttall) [E] 1840
Sparrow, Lark *Chondestes grammacus*	—	Say/Peale	—	(Say) 1823
Sparrow, Lincoln's* *Melospiza lincolnii*	—	Audubon, J.J./ Lincoln	Forbush's	(Audubon, J.J.) 1834
Sparrow, Olive *Arremonops rufivirgatus*	—	McCown	—	(Lawrence) 1851
Sparrow, Rufous-crowned *Aimophila ruficeps*	—	Heermann	Scott's; Boucard's	(Cassin) 1852
Sparrow, Rufous-winged *Aimophila carpalis*	—	Bendire [G]+	Bendire's; Bendire's Summer Finch	(Coues) 1873
Sparrow, Sage *Amphispiza belli*	—	Bell	Bell's	(Cassin) 1850
Sparrow, Savannah *Passerculus sandwichensis*	—	—	Belding's; Bryant's	(Gmelin) [G] 1789
Sparrow, Song *Melospiza melodia*	—	—	Brown's Song; Heermann's Song; Merrill's Song; Samuel's Song	(Wilson) [Sc] 1810

COMMON AND SCIENTIFIC NAMES	FIRST OBSERVER	COLLECTOR	EARLIER EPONYM	DESCRIBER/DATE
Sparrow, White-crowned *Zonotrichia leucophrys*	—	—	Gambel's; Nuttall's	(Forster) [G] 1772
Sparrow, Worthen's* *Spizella wortheni**	—	Marsh	—	Ridgway 1884
Starthroat, Plain-capped *Heliomaster constantii**	—	—	Constant's	(DeLattre) [F] 1843
Stilt, Black-necked *Himantopus mexicanus*	—	—	—	(Müller) [G] 1776
Storm-Petrel, Ashy *Oceanodroma homochroa*	—	—	—	(Coues) 1864
Storm-Petrel, Black *Oceanodroma melania*	—	—	Townsend Petrel	(Bonaparte) [I] 1854
Storm-Petrel, Guadalupe *Oceanodroma macrodactyla*[1]	—	Bryant	—	Bryant 1887
Storm-Petrel, Leach's* *Oceanodroma leucorhoa*	—	—	Leach's Petrel; Beal's Petrel; Chapman Petrel; Kaeding's Petrel	(Vieillot) [F] 1818
Storm-Petrel, Least *Oceanodroma microsoma*	—	Xántus [H]+	—	(Coues) 1864
Storm-Petrel, Wilson's* *Oceanites oceanicus*	—	—	Wilson's Petrel	(Kuhl) [G] 1820

Species				
Swallow, Cliff *Hirundo pyrrhonota*	—	Forster [G]	—	(Vieillot) [F] 1817
Swallow, N. Rough-winged *Stelgifopteryx serripennis*	Sessé [Sp]/Moziño [M]	Audubon, J.J.	—	(Audubon, J.J.) 1838
Swallow, Violet-green *Tachycineta thalassina*	—	Bullock [E]	—	(Swainson) [E] 1827
Swan, Tundra *Cygnus columbianus*	Lewis/Clark	—	Bewick's	(Ord) 1815
Swift, Vaux's* *Chaetura vauxi**	—	Townsend, J.K.	—	(Townsend, J.K.) 1839
Swift, White-throated *Aeronautes saxatalis*	Woodhouse	Kennerly/Möllhausen [G]	—	(Woodhouse) 1853
Tanager, Flame-colored *Piranga bidentata*	—	—	Swainson's	Swainson [E] 1827
Tanager, Summer *Piranga rubra*	—	Catesby [E]	Cooper's	(Linnaeus) [S] 1758
Tanager, Western *Piranga ludoviciana*	Sessé [Sp]/Moziño [M]	—	Lewis/Clark	(Wilson) [Sc] 1811
Teal, Cinnamon *Anas cyanoptera*	Sessé [Sp]/Moziño [M] ?	—	—	Vieillot [F] 1816
Tern, Elegant *Sterna elegans*	—	Gambel	—	Gambel 1849

COMMON AND SCIENTIFIC NAMES	FIRST OBSERVER	COLLECTOR	EARLIER EPONYM	DESCRIBER/DATE
Tern, Forster's* *Sterna forsteri**	—	—	—	Nuttall [E] 1834
Tern, Least *Sterna antillarum*	—	—	Brown's	(Lesson) [F] 1847
Thrasher, Bendire's* *Toxostoma bendirei**	—	Bendire [G]+	Bendire's Mocking-thrush	(Coues) 1873
Thrasher, California *Toxostoma redivivum*	La Pérouse [F]	Gambel	—	(Gambel) 1845
Thrasher, Crissal *Toxostoma crissale*	—	Henry	—	Henry 1858
Thrasher, Curve-billed *Toxostoma curvirostre*	—	Bullock [E]	Palmer's	(Swainson) [E] 1827
Thrasher, Gray *Toxostoma cinereum*	—	Xántus [H]+	Mearn's	(Xántus) [H]+ 1860
Thrasher, Le Conte's* *Toxostoma lecontei**	—	Le Conte	Le Conte's Thrush; Le Conte's Mocking-thrush	Lawrence 1851
Thrasher, Long-billed *Toxostoma longirostre*	—	—	Sennett's	(Lafresnaye) [F]1838
Thrasher, Sage *Oreoscoptes montanus*	—	Townsend, J. K.	—	(Townsend, J. K.) 1837

Species		Named for	Alternate name	Authority
Thrush, Aztec / *Ridgwayia* pinicola	—	—	—	(Sclater) [E] 1859
Thrush, Hermit / *Catharus guttatus*	—	—	Audubon's Hermit	(Pallas) [G] 1811
Thrush, Swainson's* / *Catharus ustulatus*	—	—	Alma's	(Nuttall) [E] 1840
Titmouse, Bridled / *Baeolophus wollweberi**	—	Wollweber [G]	—	(Bonaparte) [I] 1850
Titmouse, Juniper / *Baeolophus griseus*	—	—	—	(Ridgway) 1882
Titmouse, Oak / *Baeolophus inornatus*	—	Gambel	—	(Gambel) 1845
Titmouse, Tufted / *Baeolophus bicolor*	—	Catesby [E]	Sennett's	(Linnaeus) [S] 1766
Towhee, Abert's* / *Pipilo aberti**	—	Abert ?	Abert's Finch	Baird 1852
Towhee, California / *Pipilo crissalis*	—	Beechey [E]	Anthony's	(Vigors) [Ir] 1839
Towhee, Canyon / *Pipilo fuscus*	—	Bullock [E]	—	Swainson [E] 1827
Towhee, Green-tailed / *Pipilo chlorurus*	—	Townsend, J. K.	Blanding's Finch	(Audubon, J. J.) 1839

COMMON AND SCIENTIFIC NAMES	FIRST OBSERVER	COLLECTOR	EARLIER EPONYM	DESCRIBER/DATE
Towhee, Socorro *Pipilo socorroensis*	—	Grayson	—	Grayson 1867
Towhee, Spotted *Pipilo maculatus*	Sessé [Sp]/Moziño [M]	—	—	Swainson [E] 1827
Trogon, Eared *Euptilotis neoxenus*	—	Floresi [I]	—	(Gould) [E] 1838
Trogon, Elegant *Trogon elegans*	—	—	—	Gould [E] 1834
Turkey, Wild *Meleagris gallopavo*	Coronado [Sp]/ Oviedo [Sp]	—	Merriam's; Gould's	Linnaeus [S] 1758
Verdin *Auriparus flaviceps*	—	—	—	(Sundevall) [S] 1850
Vireo, Bell's* *Vireo bellii**	—	Bell	Bell's Greenlet	Audubon, J. J. 1844
Vireo, Black-capped *Vireo atricapillus*	—	Woodhouse	—	Woodhouse 1852
Vireo, Cassin's* *Vireo cassinii**	—	Xántus[H]+	Cassin's Greenlet	Xántus [H]+ 1858
Vireo, Gray *Vireo vicinior*	—	Coues	—	Coues 1866

Species				
Vireo, Hutton's* *Vireo huttoni**	—	Hutton	Anthony's; Frazar's; Hutton's Greenlet; Stephens'	Cassin 1851
Vireo, Plumbeous *Vireo plumbeus*	—	Coues	—	Coues 1866
Vireo, Warbling *Vireo gilvus*	—	—	Swainson's Warbling Greenlet	(Vieillot) [F] 1808
Vireo, Yellow-green *Vireo flavoviridis*	—	—	—	(Cassin) 1851
Vulture, Black *Coragyps atratus*	Pfefferkorn [G] ?	Bartram	—	(Bechstein) [G] 1793
Vulture, Turkey *Cathartes aura*	—	—	—	(Linnaeus) [S] 1758
Warbler, Black-throated Gray *Dendroica nigrescens*	—	Townsend, J. K.	—	(Townsend, J. K.) 1837
Warbler, Canada *Wilsonia* canadensis*	—	—	—	(Linnaeus) [S] 1766
Warbler, Colima *Vermivora crissalis*	—	Lloyd/Richardson, W. B.	—	(Salvin [E]/Godman) [E] 1889
Warbler, Golden-cheeked *Dendroica chrysoparia*	—	Salvin [E]	—	Sclater [E]/Salvin [E] 1860
Warbler, Golden-crowned *Basileuterus culicivorus*	—	Deppe [G]	Brasher's	(Deppe) [G] 1830

COMMON AND SCIENTIFIC NAMES	FIRST OBSERVER	COLLECTOR	EARLIER EPONYM	DESCRIBER/DATE
Warbler, Grace's* *Dendroica graciae*	—	Coues	—	Baird 1865
Warbler, Hermit *Dendroica occidentalis*	—	Nuttall [E]	—	(Townsend, J. K.) 1837
Warbler, Hooded *Wilsonia* citrina*	—	—	—	(Boddaert) [?] 1783
Warbler, Lucy's* *Vermivora luciae*	—	Cooper	—	(Cooper) 1861
Warbler, MacGillivray's* *Oporornis tolmie*	—	Townsend, J. K.	Tolmie	(Townsend, J. K.) 1839
Warbler, Olive *Peucedramus taeniatus*	—	—	—	(Du Bus de Gisignies) [B] 1847
Warbler, Orange-crowned *Vermivora celata*	—	Say/Peale	—	Say/Peale
Warbler, Red-faced *Cardellina rubrifrons*	—	—	—	(Giraud) 1841
Warbler, Socorro *Parula graysoni*	—	Grayson	—	(Ridgway) 1887
Warbler, Swainson's* *Limnothlypis swainsonii*	—	Bachman	—	(Audubon, J. J.) 1834

Warbler, Townsend's* *Dendroica townsendi**	—	Townsend, J. K.	—	(Townsend, J. K.) 1837
Warbler, Virginia's* *Vermivora virginiae**	—	Anderson	—	(Baird) 1860
Warbler, Wilson's* *Wilsonia* pusilla*	—	—	—	(Wilson) [Sc] 1811
Warbler, Yellow-rumped *Dendroica coronata*	—	Edwards [E]	Audubon's	(Linnaeus) [S] 1766
Waterthrush, Northern *Seiurus noveboracensis*	—	—	Grinnell's	(Gmelin) [G] 1789
Whippoorwill *Caprimulgus vociferus*	—	—	Stephens'	Wilson [Sc] 1812
Woodpecker, Acorn *Melanerpes formicivorus*	—	Bullock [E]	*bairdi*; Mearns'	(Swainson) [E] 1827
Woodpecker, Downy *Picoides pubescens*	—	Catesby [E]	Batchelder's; Gairdner's	(Linnaeus) [S] 1766
Woodpecker, Gila *Melanerpes uropygialis*	—	Kennerly/ Möllhausen [G]	Brewster's	(Baird) 1854
Woodpecker, Golden-fronted *Melanerpes aurifrons*	—	—	—	(Wagler) [G] 1829
Woodpecker, Hairy *Picoides villosus*	—	Catesby [E]	Cabanis'; Harris'	(Linnaeus) [S] 1766

COMMON AND SCIENTIFIC NAMES	FIRST OBSERVER	COLLECTOR	EARLIER EPONYM	DESCRIBER/DATE
Woodpecker, Imperial *Campephilus imperialis* [1]	—	Floresi [I]	—	(Gould) [E] 1832
Woodpecker, Ladder-backed *Picoides scalaris*	—	Deppe [G] ?	—	(Wagler) [G] 1829
Woodpecker, Lewis's* *Melanerpes lewis**	—	Lewis/ Clark	—	(Gray) [E] 1849
Woodpecker, Nuttall's* *Picoides nuttallii**	—	Gambel	—	(Gambel) 1843
Woodpecker, Strickland's* *Picoides stricklandi**	—	Mann [?]	—	(Malherbe) [F] 1845
Woodpecker, White-headed *Picoides albolarvatus*	—	Bell	—	(Cassin) 1850
Wood-Pewee, Western *Contopus sordidulus*	—	—	—	Sclater [E] 1859
Wren, Bewick's* *Thryomanes bewickii**	—	Audubon, J. J.	Baird's; Vigors'	(Audubon, J. J.) 1827
Wren, Cactus *Campylorhynchus brunneicapillus*	—	—	Bryant's	(Lafresnaye) [F] 1835
Wren, Canyon *Catherpes mexicanus*	—	—	—	(Swainson) [E] 1829

Wren, Carolina *Thryothorus ludovicianus*	—	—	Berlandier's	(Latham) [E] 1790
Wren, House *Troglodytes aedon*	Sessé [Sp]/Moziño [M]	—	—	Vieillot [F] 1809
Wren, Rock *Salpinctes obsoletus*	Sessé [Sp]/Moziño [M]	Say/Peale	—	(Say) 1823
Wren, Socorro *Thryomanes sissonii*	—	Grayson	—	(Grayson) 1868
Wrentit *Chamaea fasciata*	—	Gambel	—	(Gambel) 1845
Yellowthroat, Belding's* *Geothlypis beldingi**	—	Belding	Goldman's	Ridgway 1882

[1] Extinct
[2] Extirpated from wild

REFERENCES

Abbott, C. G. 1933. Closing History of the Guadalupe Caracara. *Condor* 35:10–14.

Abert, J. W. 1847. [Scaled Quail]. *Proc. Acad. Nat. Sci. Phil.* 3:321–322.

———. 1848. Report. In *Notes of a Military Reconnaissance, from Fort Leavenworth, in Missouri, to San Diego, in California, including Part of the Arkansas, Del Norte, and Gila Rivers by Lieut. Col. W. H. Emory.* 30th. Cong., 1st sess., Ex. Doc. 41, 419–546.

———. 1966. *The Original Travel Diary of Lieutenant J. W. Abert.* Ed. J. Galvin. San Francisco: John Howell-Books.

Aiken, C.E.H. 1937. Birds of the Southwest. Ed. E. R. Warren. Gen. Series No. 212. Colorado Springs: Colorado College Publ.

Ainley, D. G., and T. J. Lewis. 1974. The History of Farallon Island Marine Bird Populations, 1854–1972. *Condor* 76:432–446.

Alden, R. H., and J. D. Ifft. 1943. Early Naturalists in the Far West. Occ. Papers 20:1–59. San Francisco: California Academy of Science.

Aldrich, J. W. 1968. In Memoriam: Harry Church Oberholser *Auk* 85:24–29.

Aldrich, J. W., and K. P. Baer. 1970. Status and Speciation in the Mexican Duck *(Anas diaz).* *Wilson* 82:63–73.

Allen, J. A. 1879. Coues's Bibliography of Ornithology. *Bull. Nutt. Ornith. Club.* 4:175–178.

———. 1893. The Geographical Origin and Distribution of North American Birds, Considered in Relation to Faunal Areas of North America. *Auk* 10:97–150.

———. 1901. In Memoriam: George Burrill Sennett. *Auk* 18:11–23.

American Ornithologists' Union. *Check-List of North American Birds.* 1st ed. 1886; 2nd ed. 1895; 3rd ed. 1910; 4th ed. 1931; 5th ed. 1957; 6th ed. 1983; 7th ed. 1998, w/ 1999 minor errors update; including amendments through 42nd Suppl. of 2000. Lawerence, Kans.: Allen Press.

Anderson, D. W., J. E. Mandoza, and J. O. Keith. 1976. Seabirds in the Gulf of California: A Vulnerable, International Resource. *Nat. Res. Jour.* 16:483–505.

Anthony, A. W. 1892. Birds of Southwestern New Mexico. *Auk* 9:357–369.

———. 1895. The Fulmars of Southern California. *Auk* 12:100–109.

———. 1898. Petrels of Southern California. *Auk* 15:140–144.

———. 1899. A Night at Sea. *Bull. Cooper Ornith. Club* 1:101–102.

———. 1900a. Nesting Habits of the Pacific Coast Species of the Genus Puffinus. *Auk* 17:247–252.

———. 1900b. A Night on Land. *Condor* 2:28–29.

———. 1906. Random Notes on Pacific Gulls. *Auk* 13:129–137.

———. 1924. The Raided Rookeries of Laysan, a Belated Echo. *Condor* 26:33–34.

Attwater, H. P. 1892a. List of Birds Observed in the Vicinity of San Antonio, Bexar County, Texas. *Auk* 9:229–238, 337–345.

———. 1892b. Warblers Destroyed by a 'Norther.' *Auk* 9:303.

Audubon, J. J. 1827–1838. *The Birds of America.* London: J. J. Audubon.

———. 1831–1839. *Ornithological Biography.* 5 vols. Edinburgh: Adam & Charles Black.

Audubon, J. W. 1906. *Audubon's Western Journal: 1849–1850.* Cleveland: Arthur H. Clark Company.

Audubon, M. R., ed. 1960. *Audubon and His Journals.* 2 vols. New York: Dover.

Bahre, C. 1983. Human Impact: The Midriff Islands. In *Island Biogeography in the Sea of Cortez,* ed. T. J. Case and M. L. Cody, 290–306. Berkeley: Univ. Calif. Press.

Bahre, C. J. 1991. *A Legacy of Change: Historic Human Impact on Vegetation in the Arizona Borderlands.* Tucson: Univ. Ariz. Press.

Bailey, F. M. 1902. *Handbook of Birds of the Western United States.* Boston: Houghton Mifflin.

———. 1923. Birds Recorded from the Santa Rita Mountains in Southern Arizona. *Pac. Coast Avifauna* No. 15: 1–60.

———. 1928. *Birds of New Mexico.* Sante Fe: New Mex. Dept. Game & Fish.

Bailey, V. 1923. Maximilian's Travels in the Interior of North America, 1832 to 1834. *Amer. Mus. Nat. Hist.* 23:337–343.

———. 1941. Alfred Webster Anthony. *Auk* 58:439–443.

Baird, S. F. 1852. Birds. In *App. C., Exploration and Survey of the Valley of the Great Salt Lake of Utah by H. Stansbury.* Spec. sess., Senate Ex. 3.

———. 1854. Descriptions of New Birds Collected between Albuquerque, N. M., and San Francisco, California, During the Winter of 1853–54, by Dr. C.B.R. Kennerly and H. B. Möllhausen, Naturalists Attached to the Survey of the Pacific R. R. Route, under Lt. A. W. Whipple. *Proc. Acad. Nat. Sci. Phil.* 7:118–120.

———. 1859a. Birds of the Boundary. In *Report on the United States and Mexican Boundary Survey by W. H. Emory.* Vol. 2, pt. 2. 34th Cong., 1st sess., H. Ex. Doc. 135.

———. 1859b. Notes on a Collection of Birds Made by M. John Xantus, at Cape St. Lucas, Lower California, and Now in the Museum of the Smithsonian Institution. *Proc. Acad. Nat. Sci. Phil.* 11:299–306.

———. 1860. List of Birds Collected on the Colorado Expedition. In *Pt. 5, Zoology.*

Report Upon the Colorado River of the West . . . by Lt. J. C. Ives. 1861. 36th Cong., 1st sess., S. Ex. Doc.

———. 1864–1866. Review of American Birds. *Smithsonian Inst. Misc. Coll.* 181.

Baird, S. F., J. Cassin, and G. N. Lawrence. 1858. Birds. In *Reports of Explorations and Surveys, to Ascertain the Most Practicable and Economical Route for a Railroad from the Mississippi River to the Pacific Ocean.* 9, Pt. 2. 33d Cong. 2d sess., H. Ex. Doc. 91.

———. 1860. *The Birds of North America.* 2 vols. Phil.: Lippincott.

Baird, S. F., T. M. Brewer, and R. Ridgway. 1874. *A History of North American Birds.* 3 vols. Boston: Little, Brown.

———. 1884. *The Water Birds of North America.* 2 vols. Boston: Little, Brown.

Baird, S. F., and R. Ridgway. 1873. On Some New Forms of American Birds. *Bull. Essex. Inst.* 5, no. 12:197–201.

Bangs, O. 1925. The History and Characters of *Vermivora Crissalis* (Salvin and Godman). *Auk* 42:251–253.

Banks, R. C. 1987. *Obsolete English Names of North American Birds and Their Modern Equivalents.* Albuquerque: Southwest Nat. Cult. Herit. Ass.

Beck, R. H. 1899a. Additional Notes on the Birds of Santa Cruz Island, Cal. *Bull. Cooper Ornith. Club* 1:85–86.

———. 1899b. Nesting of the Santa Cruz Jay. *Bull. Cooper Ornith. Club* 1:6.

Beckham, C. W. 1887. Observations on the Birds of Southwest Texas. *Proc. U.S. Natl. Mus.* 10:633–696.

Belding, L. 1883a. Catalogue of a Collection of Birds Made Near the Southern Extremity of the Peninsula of Lower California. *Proc. U.S. Natl. Mus.* 5:532–549.

———. 1883b. Catalogue of a Collection of Birds Made Near the Southern Extremity of the Peninsula of Lower California. *Proc. U.S. Natl. Mus.* 6:344–352.

———. 1900. A Part of My Experience in Collecting. *Condor* 2:1–5.

Bendire, C. E. 1887. Notes on a Collection of Birds' Nests and Eggs from Southern Arizona. *Proc. U.S. Natl. Mus.* 10:551–558.

———. 1891. Instructions for Collecting, Preparing and Preserving Birds Eggs and Nests. *U.S. Natl. Mus. Bull.* 39, Pt. D.

———. 1892, 1895. *Life Histories of North American Birds.* 2 vols. U.S. Natl. Mus. Spec. Bull. No. 1.

Bent, A. C. 1919–1958. Life Histories of North American [Birds. . . .]. *U.S. Nat. Mus. Bull.* 107, 113, 121, 126, 130, 135, 142, 146, 162, 167, 170, 174, 176, 179, 191, 195, 196, 197, 203, and 211.

———. 1930. Francis Cottle Willard. *Auk* 47:455–456.

Bent, A. C. et al. 1968. Life Histories of North American Cardinals, Buntings, Towhees, Finches, Sparrows, and Allies. Ed. and comp. O. L. Austin, Jr. *U.S. Nat. Mus. Bull.* 237, 3 parts.

Bergtold, W. H. 1906. Concerning the Thick-Billed Parrot. *Auk* 23:425–428.

Bigelow, J. M. 1856. Report of the Botany of the Expedition. In *Reports of Explorations and Surveys . . . for a Railroad . . . to the Pacific Ocean by Lt. A. W. Whipple.* Vol. 4, Pt. 5, No. 1. 33d Cong., 2d sess., H. Ex. Doc. 91, 1–16.

Birtwell, F. J. 1901. Nesting Habits of the Evening Grosbeak *(Coccothraustes Vespertinus). Auk* 18:388–391.

Birtwell, O. M. 1901. Francis J. Birtwell. *Auk* 18:413–414.

Bishop, L. B. 1929. In Memoriam: Leverett Mills Loomis. *Auk* 46:1–13.

Blake, E. W., Jr. 1887. Summer Birds of Santa Cruz Island, California. *Auk* 4:328–330.

Bolton, H. E., ed. 1916. *Spanish Explorations in the Southwest, 1542 to 1706.* New York: Charles Scribner's Sons.

Bonaparte, C. [L.] 1825. Descriptions of Two New Species of Mexican Birds. *J. Acad. Nat. Sci. Phil.* 4:387–390.

Bonaparte, C. L. 1825–1833. *American Ornithology,* or *The National History of Birds Inhabiting the United States.* 4 vols. Phil.: Carey, Lea, & Carey.

Bonaparte, C. [L.] 1837. The Prince of Musignano . . . Descriptions of New or Interesting Birds from Mexico and South American. *Proc. Zool. Soc. London* Pt. 5, 108–114.

Bourke, J. G. [1891] 1971. *On the Border with Crook.* Reprint, Lincoln: Univ. Nebraska Press.

Brandt, H. W. 1951. *Arizona and Its Bird Life.* Cleveland: Bird Research Foundation.

Breninger, G. F. 1898. The Ferruginous Pygmy Owl. *Osprey* 2:128.

———. 1899. A Nest of the Blue-Throated Hummingbird. *Osprey* 3:86–87.

———. 1904. San Clemente Island Birds. *Auk* 21:218–223.

Brewster, W. 1876. Ornithology of the Wheeler Expeditions. *Bull. Nutt. Ornith. Club* 1:70–71.

———. 1879a. Notes upon the Distribution, Habits, and Nesting of the Black-Capped Vireo *(Vireo Atricapillus). Bull. Nutt. Ornith. Club* 4:99–103.

———. 1879b. On the Habits and Nesting of Certain Rare Birds in Texas. *Bull. Nutt. Ornith. Club* 4:75–80.

———. 1881. Additions to the Avi-Fauna of the United States. *Bull. Nutt. Ornith. Club* 6:252.

———. 1882. On a Collection of Birds Lately Made by Mr. F. Stephens in Arizona. *Bull. Nutt. Ornith. Club* 7:65–86, 135–147, 193–212.

———. 1883. On a Collection of Birds Lately Made by Mr. F. Stephens in Arizona. *Bull. Nutt. Ornith. Club* 8:21–36.

———. 1885. Preliminary Notes on Some Birds Obtained in Arizona by Mr. F. Stephens in 1884. *Auk* 2:84–85.

———. 1887. Further Notes on the Masked Bob-white. . . . *Auk* 4:159–160.

———. 1889. Descriptions of Supposed New Birds from Western North America. *Auk* 6:85–98.

———. 1895. Apparatus for Preparing Birds' Eggs. *Auk* 12:196–97.

———. 1898. Occurrence of the Spotted Screech Owl *(Megasops Aspersus)* in Arizona. *Auk* 15:186.

———. 1902. Birds of the Cape Region of Lower California. *Bull. Mus. Comp. Zool.* 42, no. 1:1–241.

———. 1910. In Memoriam: James Cushing Merrill. *Auk* 27:112–119.

Brewster, W., J. A. Allen, and E. Coues. 1883. (AOU). *Bull. Nutt. Ornith. Club* 8:221.

Brodhead, M. J. 1973. A Soldier-Scientist in the American Southwest. *Hist. Mono.* No. 1. Tucson: Ariz. Hist. Soc.

Brooks, A. 1922. What Color Are the Feet of the Western Gull? *Condor* 24:94–95.

Brooks, M. 1938. Allen Brooks—A Biography. *Condor* 40:12–17.

Brown, B. T., S. W. Carothers, and R. R. Johnson. 1987. *Grand Canyon Birds.* Tucson: Univ. Ariz. Press.

Brown, H. 1899a. The California Vulture in Arizona. *Auk* 16:272.

———. 1899b. Field Notes. Ed. T. Huals, Tucson: Univ. Ariz. Bird Mus.

———. 1899c. The Scarlet Ibis *(Guara Rubra)* in Arizona. *Auk* 16:270.

———. 1900. The Conditions Governing Bird Life in Arizona. *Auk* 17:31–34.

———. 1901a. A Band-Tailed Hawk's Nest. *Auk* 18:392–393.

———. 1901b. Bendire's Thrasher. *Auk* 18:225–231.

———. 1903. Arizona Bird Notes. *Auk* 20:43–50.

———. 1904. Masked Bob-White *(Colinus ridgway).* *Auk* 21:209–213.

———. 1906. The Water Turkey and Tree Ducks near Tucson, Arizona. *Auk* 23:217–218.

Brown, N. C. 1882. A Reconnaissance in Southwestern Texas. *Bull. Nutt. Ornith. Club* 7:33–42.

———. 1884. A Second Season in Texas. *Auk* 1:120–124.

Bryant, W. E. 1887. Additions to the Ornithology of Guadalupe Island. *Bull. Cal. Acad. Sci.* 2:269–318.

———. 1888. Birds and Eggs from the Farallon Islands. *Proc. Calif. Acad. Sci.* 2d ser. 1:25–50.

———. 1889. A Catalogue of the Birds of Lower California, Mexico. *Proc. Calif. Acad. Sci.* 2d ser. 2:237–320.

———. 1891a. Andrew Jackson Grayson. *Zoe* 2:34–68.

———. 1891b. The Cape Region of Baja California. *Zoe* 2:185–201.

Burns, F. L. 1917. Miss Lawson's Recollections of Ornithologists. *Auk* 34:275–282.

———. 1932. Charles W. and Titian R. Peale and the Ornithological Section of the Old Philadelphia Museum. *Wilson Bull.* 44.

Butcher, H. B. 1868. List of Birds Collected at Laredo, Texas, in 1866 and 1867. *Proc. Acad. Nat. Sci. Phil.* 20:148–150.

Cabeza de Vaca, A. N. 1993. *The Account: Álvar Núñez Cabeza de Vaca's Relación.* Trans. M. A. Favata and J. B. Fernández. Houston, Texas: Arte Público Press.

Carothers, S. W., and R. R. Johnson. 1976. The Mississippi Kite in Arizona: A Second Record. *Condor* 78:114–115.

Casanova, F. E., ed. 1968. General Crook Visits the Supais: As Reported by John G. Bourke. *Ariz. and the West* 10:253–276.

Cassin, J. 1850. Descriptions of New Species of Birds. . . . *Proc. Acad. Nat. Sci. Phil.* 5:103–106.

———. 1851. Sketch of the Birds . . . Previously Known and Descriptions of Three New Species. *Proc. Acad. Nat. Sci. Phil.* 5:149–154.

———. 1852. Descriptions of New Species of Birds, Specimens of Which Are in the Collection of the Academy of Natural Sciences of Philadelphia. *Proc. Acad. Nat. Sci. Phil.* 6:184–188.

———. 1856. *Illustrations of the Birds of California, Texas, Oregon, British and Russian America.* Phil.: Lippincott.

———. 1865. Notes on Some New and Little Known Rapacious Birds. *Proc. Acad. Nat. Sci. Phil.* 17:17.

Catesby, M. 1731–1743. *The Natural History of Carolina, Florida, and the Bahama Islands.* Vols. 1 and 2. London.

Chapman, F. M. 1891. On the Birds Observed near Corpus Christi, Texas, During Parts of March and April, 1891. *Bull. Amer. Mus. Nat. Hist.* 3:315–328.

———. 1894. Remarks on the Origin of Bird Migration. *Auk* 11:12–17.

———. 1916. Florence Merriam Bailey. *Bird-Lore* 18:142–144.

———. 1928. In Memoriam: Louis Agassiz Fuertes. *Auk* 45:1–26.

Choate, E. A. 1985. *The Dictionary of American Bird Names.* Rev. ed. R. A. Paynter, Jr. Boston: Harvard Common Press.

Coale, H. K. 1894. Ornithological Notes on a Flying Trip through Kansas, New Mexico, Arizona and Texas. *Auk* 11:215–222.

Coan, E. 1981. *James Graham Cooper: Pioneer Western Naturalist.* Moscow: Univ. Press Idaho.

Colton, H. S. 1930. A Brief Survey of the Early Expeditions into Northern Arizona. *Mus. N. Ariz. Mus. Notes* 2, no. 9:1–4.

———. 1932. Samuel Washington Woodhouse. *Mus. N. Ariz. Mus. Notes* 5, no. 1:1–4.

Cooke, W. W. 1916. The Type Locality of *Brachyramphus craverii*. *Auk* 33:80.

Cooper, J. G. 1861. New Californian Animals. *Proc. Calif. Acad. Sci.* 2:118–123.

———. 1869. The Naturalist in California. *Amer. Naturalist* 3:182–189, 470–481.

———. [1870] 1974. *Ornithology.* Ed. S. F. Baird. Vol. 1. Land Birds. Geol. Sur. Calif. Reprint, New York: Arno Press.

Couch, D. N. 1854. Descriptions of New Birds of Northern Mexico. *Proc. Acad. Nat. Sci. Phil.* 7:66–67.

Coues, E. 1861. A Monograph of the Tringeae of North America. *Proc. Acad. Nat. Sci. Phil.* 13:170–205.

———. 1865. Ornithology of a Prairie-Journey, and Notes on the Birds of Arizona. *Ibis* 2d ser. 1:157–165.

———. 1866a. From Arizona to the Pacific. *Ibis* 2d ser. 2:259–75.

———. 1866b. List of the Birds of Fort Whipple, Arizona, with Which Are Incorporated All Other Species Ascertained to Inhabit the Territory; with Brief Critical and Field Notes, Descriptions of New Species, Etc. *Proc. Acad. Nat. Sci. Phil.* 18:39–100.

———. 1868. List of Birds Collected in Southern Arizona by Dr. E. Palmer; with Remarks. *Proc. Acad. Nat. Sci. Phil.* 20:81–85.

———. 1872. *Key to North American Birds*. Salem: Naturalists' Agency.

———. 1874a. *Birds of the Northwest: A Handbook of the Ornithology of the Region Drained by the Missouri River and Its Tributaries*. Misc. Publ. No. 3. U.S. Geol. Sur. Terr. New York: Arno Press.

———. 1874b. *Field Ornithology*. Salem: Naturalists' Agency.

———. 1875. An Account of the Various Publications [and Zoology] Relating to the Travels of Lewis and Clarke. . . . *Bull. U. S. Geol. & Geog. Sur. Terr.* 1, bull. 6:417–444.

———. 1878. *Birds of the Colorado Valley*. Misc. Publ. No. 11. U.S. Geol. Sur. Terr.

———. 1879a. Note on the Black-Caped Greenlet, *Vireo Atricapillus* of Woodhouse. *Bull. Nutt. Ornith. Club* 4:192–194.

———. 1879b. Second Installment of American Ornithological Bibliography. *Bull. U.S. Geol. & Geog. Sur. Terr.* 5, bull. 2:239–330.

———. 1879c. Third Installment of American Ornithological Bibliography. *Bull. U.S. Geol. & Geog. Sur. Terr.* 5, bull. 4:521–1006.

———. 1897. Dr. Coues' Column. *Osprey* 1:113.

———., ed. 1893. *History of the Expedition under the Command of Lewis and Clark*. 3 vols. New York: Dover.

Curtis, C. A. 1902. Coues at His First Army Post. *Bird-Lore* 4:5–9.

Cutright, P. R. 1969. *Lewis & Clark: Pioneering Naturalists*. Urbana: Univ. Ill. Press.

Cutright, P. R., and M. J. Brodhead. 1981. *Elliott Coues: Naturalist and Frontier Historian*. Chicago: Univ. Ill. Press.

Dall, W. H. 1915. *Spencer Fullerton Baird*. Phil.: Lippincott.

Dana, R. H. 1964. *Two Years Before the Mast*. New York: Penguin Putnam.

Darwin, C. [1839–1843] 1962. *The Voyage of the Beagle*. Garden City, N.Y.: Doubleday.

———. [1859] 1964. *On the Origin of Species*. Reprint, Cambridge, Mass.: Harvard Univ. Press.

Davis, J. 1951. Distribution and Variation of the Brown Towhees. *Univ. Calif. Publ. Zool.* 52:1–120.

Dawson, W. L. 1913. Allan Brooks—An Appreciation. *Condor* 15:69–76.

———. 1923. *Birds of California*. 4 vols. San Diego: South Moulton.

Deane, R. 1905. A Hitherto Unpublished Letter of John James Audubon. *Auk* 22:172–175.

DeSante, D. F., and D. G. Ainley. 1980. The Avifauna of the South Farallon Islands, California. *Studies in Avian Biol.* 4.

Dixon, J. B. 1937. The Golden Eagle in San Diego County, California. *Condor* 39:49–56.

Douglas, D. 1829. Observations on the *Vultur Californianus* of Shaw. *Zool. Journal* 4:328–330.

Dresser, H. E. 1865. Notes on the Birds of Southern Texas. *Ibis* 2d ser. 1:312–330, 466–495.

———. 1866. Notes on the Birds of Southern Texas. *Ibis* 2d ser. 2:23–46.

Dunlap, E. 1988. Layson Albatross Nesting on Guadalupe Island, Mexico. *Amer. Birds* 42:180–181.

Dutcher, W. 1904. Report of the National Association of Audubon Societies. *Bird-Lore.*

Dwight, J. 1919. Description of a New Race of the Western Gull. *Biol. Soc. Wash.* 32:11–13.

Elliot, D. G. 1896. In Memoriam: George Newbold Lawrence. *Auk* 13:1–10.

———. 1901. In Memoriam: Elliott Coues. *Auk* 18:1–11.

———. 1902. Coues As a Young Man. *Bird-Lore* 4:3–5.

Emerson, W. O. 1899. Dr. James G. Cooper. *Bull. Cooper Ornith. Club* 1:1–5.

———. 1902. In Memoriam: Dr. James G. Cooper. *Condor* 4:101–103.

———. 1904. The Farallone Islands Revisited, 1887–1903. *Condor* 6:61–68.

Emory, W. H. 1848. *Notes on a Military Reconnaissance, from Fort Leavenworth, in Missouri, to San Diego, California, Including Part of the Arkansas, Del Norte, and Gila River.* 30th Cong., 1st sess., Ex. Doc. 41.

———. 1857, 1859. *Report on the United States and Mexican Boundary Survey.* 3 vols. 34th Cong., 1st sess., H. Ex. Doc. 135.

Engstrand, H. W. 1981. *Spanish Scientists in the New World.* Seattle: Univ. Wash. Press.

Ewan, J., and N. D. Ewan. 1981. *Biographical Dictionary of Rocky Mountain Naturalists.* Boston: Dr. W. Junk.

Farquhar, F. P. 1925. Colonel Benson. *Sierra Club Bull.* 12:174–179.

Finley, W. L. 1906, 1908. Life History of the California Condor. Pts. 1–3. *Condor* 8:134–142; 10:4–10, 58–65.

Finley, W. L., and I. Finley. 1915. With the Field-Agents, Bird-Friends in Arizona. *Bird-Lore* 17:237–245.

Fisher, A. K. 1894. The Capture of *Basilinna Leucotis* in Southern Arizona. *Auk* 11:325–326.

———. 1920. In Memoriam: Lyman Belding. *Auk* 37:32–45.

Fisher, W. K. 1903. Dr. Edgar A. Mearns. *Condor* 5:109.

———. 1905. In Memoriam: Walter E. Bryant. *Condor* 7:129–131.

———. 1918. In Memorian: Lyman Belding. *Condor* 20:50–61.

———. 1923. William Wrightman Price. *Condor* 25:50–57.

Fleming, T. 1983. They've Found the Missing Masterpieces. *International Wildlife* 13:19–24.

Ford, F. 1911. Preliminary List of Birds of New Mexico. In *Rpt. No. 1.* Cons. Nat. Res. Comm. New Mexico. 17–63.

Fowler, F. H. 1903. Stray Notes from Southern Arizona. *Condor* 5:68–71, 106–107.

Frazar, A. M. 1881. Destruction of Birds by Storm while Migrating. *Bull. Nutt. Ornith. Club* 6:249–252.

Friedmann, H., L. Griscom, and R. T. Moore. 1950. Distributional Check-List of the Birds of Mexico. Pt. 1. *Pacific Coast Avifauna* 29.

Fuertes, L. A. 1903. With the Mearns Quail in Southwestern Texas. *Condor* 5:113–116.

Gambel, W. 1843. Description of Some New and Rare Birds of the Rocky Mountains and California. *Proc. Acad. Nat. Sci. Phil.* 1:258–262.

———. 1845. Descriptions of New and Little Known Birds Collected in Upper California. *Proc. Acad. Nat. Sci. Phil.* 2:263–266.

———. 1846. Remarks on the Birds in Upper California. *Proc. Acad. Nat. Sci. Phil.* 3:44–48, 110–115.

———. 1847. Remarks on the Birds in Upper California. *Proc. Acad. Nat. Sci. Phil.* 3:154–158, 200–205.

Gaut, J. H. 1904. [Field Notes] In *Guide to the Field Reports of the United States Fish and Wildlife Service. Circa 1860–1961 by W. E. Cox.* Series 3, Mexico: Chihuahua, Box 124, Folder 23, Spec. Rpts. No. 4. Arch. and Spec. Coll. Smithsonian Inst.

Gaylord, H. A. 1897. Notes from Guadalupe Island. *Nidologist* 4:41–43.

———. 1899. Spring Migration of 1896 in the San Gabriel Valley. *Bull. Cooper Ornith. Club* 1:7–8.

Geiser, S. W. 1937. *Naturalists of the Frontier.* Dallas: So. Methodist Univ.

Gilman, M. F. 1909. Among the Thrashers in Arizona. *Condor* 11:49–50.

Giraud, J. P., Jr. 1841. A Description of Sixteen New Species of North American Birds. *Ann. Lyc. Nat. Hist. N.Y.* Vol. 1.

Goetzmann, W. H. 1991. *Army Exploration in the American West, 1803–1863.* Austin: Texas State Hist. Assoc.

Goldman, E. A. 1935. Edward William Nelson—Naturalist. *Auk* 52:135–148.

Goss, N. S. 1884. *Brachyramphus Hypoleucus* off the Coast of Southern California. *Auk* 1:396.

———. 1888. New and Rare Birds Found Breeding on the San Pedro Martir Isle. *Auk* 5:240–244.

Grinnell, J. 1902a. Check-List of California Birds. *Pac. Coast Avifauna* 3.

———. 1902b. The Ornithological Writings of Dr. J. G. Cooper. *Condor* 4:103–105.

———. 1905. The Ornithological Writings of Walter E. Bryant. *Condor* 7:131–132.

———. 1908. Birds of a Voyage on Salton Sea. *Condor* 10:185–191.

———. 1909. The First Zone-tailed Hawk in California. *Condor* 11:69.

———. 1926. The Evidence as to the Former Breeding of the Rhinoceros Auklet in California. *Condor* 28:37–40.

Griscom, L., and M. S. Crosby. 1925. Birds of the Brownsville Region, Southern Texas. *Auk* 42:432–440, 519–537.

———. 1926. Birds of the Brownsville Region, Southern Texas. *Auk* 43:18–36.

Grosvenor, G., and A. Wetmore, eds. 1932–1937. *The Book of Birds.* 2 vols. Washington, D.C.: Natl. Geog. Soc.

Gustafson, A. M. 1966. *John Spring's Arizona.* Tucson: Univ. Ariz. Press.

Hancock, J. L. 1887. Notes and Observations on the Ornithology of Corpus Christi and Vicinity, Texas. *Bull. Ridgway Ornith. Club* 2:11–23.

Hanna, W. C. 1909. The White-Throated Swift on Slover Mountain. *Condor* 11:77–81.

———. 1917. Further Notes on the White-Throated Swifts of Slover Mountain. *Condor* 10:1–8.

Hargrave, L. L. 1939. Bird Bones from Abandoned Indian Dwellings in Arizona and Utah. *Condor* 41:206–210.

———. 1970. Mexican Macaws: Comparative Osteology and Survey Remains from the Southwest. *Anthro. Papers Univ. Ariz.* No. 20., Tucson: Univ. Arizona Press.

Harris, H. 1934. Notes on the Xántus Tradition. *Condor* 36:191–201.

———. 1941. The Annals of Gymnogyps to 1900. *Condor* 43:3–55.

———. 1946. An Appreciation of Allan Brooks, Zoological Artist: 1869–1846. *Condor* 48:145–153.

Hasbrouck, E. M. 1891a. The Carolina Paroquet *(Conurus Carolinensis)*. *Auk* 8:368–379.

———. 1891b. The Present Status of the Ivory-Billed Woodpecker *(Campephilus Principalis)*. *Auk* 8:174–186.

Hector, D. P. 1980. Our Rare Falcon of the Desert Grassland. *Birding* 12:92–102.

Heermann, A. L. 1853a. Catalogue of the Oological Collection in the Academy of Natural Sciences of Philadelphia. *Proc. Acad. Nat. Sci. Phil.* 6:1–36.

———. 1853b. Notes on the Birds of California, Observed During a Residence of Three Years in That Country. *J. Acad. Nat. Sci. Phil.* 2d ser. 2:259–272.

———. 1859a. Report upon Birds Collected on the Survey. In *Reports of Exploration and Surveys . . . for a Railroad . . . to the Pacific Ocean by Lt. J. G. Parke.* Zoological Report. 10, no. 1:9–21. 33d Cong., 2d sess., S. Ex. Doc. 78.

———. 1859b. Report upon Birds Collected on the Survey. In *Reports of Exploration and Surveys . . . for a Railroad . . . to the Pacific Ocean by Lt. R. S. Williamson.* 10, pt. 4, no. 2:29–80. 33d Cong., 2d sess., S. Ex. Doc. 78.

Henry, T. C. 1855. Notes Derived from Observations Made on the Birds of New Mexico during the Years 1853 and 1854. *Proc. Acad. Nat. Sci. Phil.* 7:306–317.

———. 1858. Description of a New Toxostoma from Fort Thorn, New Mexico. *Proc. Acad. Nat. Sci. Phil.* 10:117–118.

———. 1859. Catalogue of the Birds of New Mexico, As Compiled from Notes and Observations Made While in That Territory, during a Residence of Six Years. *Proc. Acad. Nat. Sci. Phil.* 11:104–109.

Henshaw, H. W. 1873. Report upon Ornithological Specimens. *In* Sec. 3, 95–148. Geog. & Geol. Expl. Sur. West 100th Meridian.

———. 1874. On a Hummingbird New to Our Fauna with Certain Other Facts Ornithological. *Amer. Naturalist.* 8:241–243.

———. 1875a. Annotated List of the Birds of Arizona. *In* App. I 2 of App. LL, 153–166. Ann. Rpt. Geog. & Geol. Expl. Sur. West 100th Meridian.

———. 1875b. The Ornithological Collections. *In* vol. 5. Zoology. Chap. 3, 133–507. Geog. & Geol. Expl. Sur. West 100th Meridian.

———. 1876. Report on the Ornithology of Portions of California. *In* App. JJ, 224–278. Geog. & Geol. Expl. Sur. West 100th Meridian.

———. 1877. Description of a New Species of Humming Bird from California. *Nutt. Ornith. Club* 2:53–58.

———. 1881. On *Podiceps Occidentalis* and *P. Clarkii. Nutt. Ornith. Club* 6:211–216.

———. 1885. List of Birds Observed in Summer and Fall on the Upper Pecos River, New Mexico. *Auk* 2:326–333.

———. 1886. List of Birds Observed in Summer and Fall on the Upper Pecos River, New Mexico. *Auk* 3:73–80.

———. 1919. Autobiographical Notes. *Condor* 21:102–107, 165–171, 177–181, 217–222.

———. 1920a. Autobiographical Notes. *Condor* 22:3–10, 55–60, 95–101.

———. 1920b. In Memoriam: William Brewster. *Auk* 37:1–27.

Hoffmann, R. 1927. *Birds of the Pacific States.* Boston: Houghton Mifflin.

Holland, H. M. 1930. Francis Cottle Willard. *Oologist* 47:33–34.

Hollister, N. 1908. Birds of the Region about Needles, California. *Auk* 25:454–462.

Hood, J. M. 1933. Andrew J. Grayson: The Audubon of the Pacific. *Auk* 50:396–402.

Howard, H. 1971. In Memoriam: Loye Holmes Miller. *Auk* 88:276–285.

Howard, O. W. 1899. Some of the Summer Flycatchers of Arizona. *Bull. Cooper Ornith. Club* 1:103–107.

———. 1900. Nesting of the Mexican Wild Turkey in Huachuca Mtns., Ariz. *Condor* 2:55–57.

———. 1902. Nesting of the Prairie Falcon. *Condor* 4:57–59.

Hubbard, J. P. 1969. The Bailey-Law Collection. *Atlantic Nat.* 24, no. 4:201–203.

———. 1977. The Biological and Taxonomic Status of the Mexican Duck. *Bull. New Mexico Dept. Game Fish* 16.

Huey, L. M. 1938. Frank Stephens, Pioneer. *Condor* 40:101–110.

Hume, E. E. [1942] 1978. *Ornithologists of the United States Army Medical Corps.* Baltimore: John Hopkins Press. Reprint, New York: Arno Press.

International Boundary Commission. 1898. *Report of the Boundary Commission upon the Survey and Re-marking of the Boundary between the United States and Mexico West of the Rio Grande, 1891 to 1896.* 3 vols. Washington: Government Printing Office.

Ives, J. C. 1861. *Report upon the Colorado River of the West.* 36th Cong., 1st sess., S. Ex. Doc.

Jaeger, E. C. 1948. Does the Poor-Will Hibernate? *Condor* 50:45–46.

———. 1949. Further Observations on the Hibernation of the Poor-Will. *Condor* 51:105–109.

Johnson, R. R., and J. M. Simpson. 1996. Unpublished Notes on Early Arizona Naturalists.

Kaeding, H. B. 1903. Bird Life on the Farallone Islands. *Condor* 5:121–127.

———. 1905. Birds from the West Coast of Lower California and Adjacent Islands. *Condor* 7:105–111, 134–137.

Kennerly, C.B.R. 1856. Field Notes and Explanations. In *Reports of Explorations and Surveys . . . for a Railroad . . . to the Pacific Ocean by Lt. A. W. Whipple.* 4, pt. 6, no. 1:5–17. 33d Cong., 2d sess., H. Ex. Doc. 91.

———. 1859. Report upon the Birds of the Route. In *Reports of Explorations and Surveys . . . for a Railroad . . . to the Pacific Ocean by Lt. A. W. Whipple.* 10, pt. 6, no. 3:19–35. 33d Cong., 2d sess., S. Ex. Doc. 78.

Kofalk, H. 1989. *No Woman Tenderfoot: Florence Merriam Bailey, Pioneer Naturalist.* College Station: Texas A. & M. Univ. Press.

Kofoid, C. A. 1923. A Little Known Ornithological Journal and Its Editor, Adolphe Boucard, 1839–1904. *Condor* 25:85–89.

Lambrecht, K. 1935. In Memoriam: Robert Wilson Shufeldt, 1850–1934. *Auk* 52:358–361.

Law, J. E. 1913–1919. *Field Notes and Letters.* Blacksburg: Virginia Tech. Mus. Nat. Hist.

———. 1917. Unpublished Letters. G. Monson collection.

Lawrence, G. N. 1851. Descriptions of New Species of Birds. . . . *Ann. Lyc. Nat. Hist. N.Y.* 5:112–124.

———. 1877. Note on *Doricha Enicura. Nutt. Ornith. Club* 2:108–109.

Lawson, J. 1709. *A New Voyage to Carolina. . . .* London.

Levy, S. H. 1971. The Mississippi Kite in Arizona. *Condor* 73:476.

Ligon, J. S. 1961. *New Mexican Birds and Where to Find Them.* Albuquerque: Univ. New Mex. Press.

Linsdale, J. M. 1936. Harry Schelwald Swarth. *Condor* 38:155–168.

Lloyd, W. 1887. Birds of Tom Green and Concho Counties, Texas. *Auk* 4:181–193, 289–299.

Loomis, L. M. 1895, 1896, 1900. California Water Birds. *Proc. Cal. Acad. Sci.* 2d & 3d ser., nos. 1–7.

Lumholtz, C. 1902. *Unknown Mexico.* Vol. 1. New York: Charles Scribner's Sons.

Lusk, R. D. 1899. Nesting of the Sulphur-Bellied Flycatcher. *Bull. Cooper Ornith. Club* 1:112–113.

———. 1900. Parrots in the United States. *Condor* 2:129.

———. 1921. The White-Eared Hummingbird in the Catalina Mountains, Arizona. *Condor* 23:99.

Madden, H. M., ed. 1949. California for Hungarian Readers, Letters of Janos Xántus, 1857 and 1859. *Calif. Hist. Soc. Quart.* 28:125–142.

Mailliard, J. 1899. Spring Notes on the Birds of Santa Cruz Island, Cal., April, 1898. *Bull. Cooper Ornith. Club* 1:41–45.

———. 1900. Measurement of the Santa Cruz Jay. *Condor* 2:42.

———. 1937. In Memoriam: Harry Schelwald Swarth. *Auk* 54:127–134.

Marsh, C. H. 1884. Birds of Silver City. *Ornith. & Ool.* 9:72–74, 126–127.

Marshall, J. T., Jr. 1957. Birds of Pine-Oak Woodland in Southern Arizona and Adjacent Mexico. *Pac. Coast Avifauna* 32.

Maximilian, Prince of Wied. 1966. *Early Western Travels.* Ed. R. G. Thwaites. Vol. 23. New York: AMS Press.

McCall, G. A. 1851. Some Remarks on the Habits, &c., of Birds Met with in Western Texas, between San Antonio and the Rio Grande, and in New Mexico; with Descriptions of Several Species Believed to Have Been Hitherto Undescribed. *Proc. Acad. Nat. Sci. Phil.* 5:213–224.

McCaskie, G. 1983. Another Look at the Western and Yellow Footed Gulls. *West. Birds* 14:85–107.

McCauley, C.A.H. 1877. Notes on the Ornithology of the Region about the Source of the Red River of Texas, from Observations Made during the Exploration Conducted by Lt. E. H. Ruffner, Corps of Engineers, U.S.A. *In* Bull. U.S. Geol. & Geog. Sur. Terr. Vol. 3, bull. 3, art. 26:655–695.

McCormick, A. I. 1899. Breeding Habits of the Least Tern in Los Angeles County, California. *Bull. Cooper Ornith. Club* 1:49–50.

McCown, J. P. 1853. Facts and Observations from Notes Taken When in Texas. *Ann. Lyc. Nat. Hist. N.Y.* 6:9–14.

McKinley, D. 1964. History of the Carolina Parakeet in Its Southwestern Range. *Wilson* 76:68–93.

McVaugh, R. 1956. *Edward Palmer: Plant Explorer of the American West.* Norman: Univ. Oklahoma Press.

———. 1982. The Lost Paintings of the Sessé & Mociño Expedition: A Newly Available Resource. *Taxon* 3:691–692.

Mearns, B., and R. Mearns. 1992. *Audubon to Xántus.* London: Academic Press.

Mearns, E. A. 1886. Some Birds of Arizona. *Auk* 3:60–73.

———. 1890. Observations on the Avifauna of Portions of Arizona. *Auk* 7:45–55, 251–264.

———. 1896. Ornithological Vocabulary of the Moki Indians. *American Anthro.* 9:390–403.

———. 1907. Mammals of the Mexican Boundary of the United States. *U.S. Natl. Mus. Bull.* 56.

Merriam, C. H. 1890. Results of a Biological Survey of the San Francisco Mountain Region and Desert of the Little Colorado in Arizona. *N. Amer. Fauna* 3.

———. 1897. The Late Major Charles E. Bendire. *Oologist* 14:37–39.

———. 1916. To the Memory of John Muir. *Sierra Club Bull.* 10:146–151.

———. 1924. Baird the Naturalist. *Sci. Monthly* 18:588–595.

Merriam, F. A. 1896a. *A-Birding on a Bronco.* Boston: Houghton Mifflin.

———. 1896b. Notes on Some of the Birds of Southern California. *Auk* 13:115–124.

Merrill, J. C. 1876. Notes on Texas Birds. *Nutt. Ornith. Club* 1:88–89.

———. 1877a. A Humming-Bird New to the Fauna of the United States. *Nutt. Ornith. Club* 2:26.

———. 1877b. Notes on *Molothrus Aeneus,* Wagl. *Nutt. Ornith. Club* 2:85–87.

———. 1878. Notes on the Ornithology of Southern Texas. *Proc. U.S. Natl. Mus.* 1:118–173.

———. 1898. In Memoriam: Charles Emil Bendire. *Auk* 15:1–6.

Miller, A. H., H. Friedmann, L. Griscom, and R. T. Moore, eds. 1957. Distributional Check-List of the Birds of Mexico. Pt. 2. *Pacific Coast Avifauna* 33.

Miller, L. 1950. *Lifelong Boyhood.* Berkeley: Univ. Calif. Press.

———. 1952. *Music in Nature.* (L. P. Record) Los Angeles: Cooper Ornith. Society.

Möllhausen, B. 1969. *Diary of a Journey from the Mississippi to the Coasts of the Pacific.* 2 vols. New York: Johnson Reprint Corp.

Monson, G. 1942. Notes on Some Birds of Southeastern Arizona. *Condor* 44:222–225.

———. 1954. Westward Extension of Ranges of the Inca Dove and Bronzed Cowbird. *Condor* 56:229–230.

———. 1964. Ornithological Aspects of Merriam's 1889 Studies As Viewed 75 Years Later. *Plateau* 37:56–60.

Monson, G., and A. R. Phillips. 1981. *Annotated Checklist of the Birds of Arizona.* Tucson: Univ. Ariz. Press.

Moore, J. A. 1986. Zoology of the Pacific Railroad Surveys. *Amer. Zool.* 26:331–341.

Morcom, G. F. 1887. Notes on the Birds of Southern California and Southwestern Arizona. *Bull. Ridgway Ornith. Club* 2:36–57.

Morfi, J. A. 1935. *History of Texas, 1673–1779.* Trans. C. E. Castañeda. Albuquerque: Quivira Soc.

Morris, G. S. 1902. Edward Harris. *Cassinia* 6:1–5.

Murphy, R. C. 1950. Frank Michler Chapman, 1865–1945. *Auk* 67:307–315.

Nelson, E. W. 1898. The Imperial Ivory-billed Woodpecker, *Campephilus Imperialis* (Gould). *Auk* 15:217–223.

———. 1913a. Herbert Brown. *Auk* 30:472.

———. 1913b. Herbert Brown—A Biographical Note. *Condor* 15:186–187.

———. 1932. Henry Wetherbee Henshaw—Naturalist. *Auk* 49:398–427.

———. 1966. *Lower California and Its Natural Resources.* Riverside: Manessier Publ.

Nordhoff, C. 1874. The Farallon Islands. *Harper's New Monthly Mag.* 48:617–625.

Norris, S. M. 1997. The Royal Botanical Expedition to New Spain, 1786–1803; Ichthyological Contributions. Tempe, AZ: Ariz. State Univ. (unpublished).

Nuechterlein, G. L. 1981. Courtship Behavior and Reproductive Isolation Between Western Grebe Color Morphs. *Auk* 98:335–349.

Nuttall, T. 1832–1834. *A Manual of the Ornithology of the United States and Canada.* 2 vols. Boston: Hillard, Gray.

Oberholser, H. C. 1933. Robert Ridgway: A Memorial Appreciation. *Auk* 50:159–169.

———. 1974. *The Bird Life of Texas.* Ed. E. B. Kincaid, Jr. 2 vols. Austin: Univ. Texas Press.

Oehser, P. H. 1952. In Memoriam: Florence Merriam Bailey. *Auk* 69:19–26.

Ogilby, J. D. 1882. A Catalogue of Birds Obtained in Navarro County, Texas. *Sci. Proc. Royal Dublin Soc.* 3:169–249.

Opler, M. E. 1941. *An Apache Life-Way*. Chicago: Univ. Chicago.

Osgood, W. H. 1903. A List of Birds Observed in Cochise County, Arizona. *Condor* 5:128–131, 149–151.

Palmer, T. S. 1917a. Botta's Visit to California. *Condor* 19:159–161.

———. 1917b. In Memoriam: Wells Woodbridge Cooke. *Auk* 34:118–132.

———. 1918. Costa's Hummingbird—Its Type Locality, Early History, and Name. *Condor* 20:114–116.

———. 1921. In Memoriam: William Dutcher. *Auk* 38:501–513.

———. 1926. Marston Abbott Frazar. *Auk* 43:579–580.

———. 1928a. The Forty-Fifth Stated Meeting of the American Ornithologist's Union. *Auk* 45:70–81.

———. 1928b. Notes on Persons Whose Names Appear in the Nomenclature of California Birds. *Condor* 30:261–307.

———. 1931. The Scientific Name of the Western Sandpiper—Who Was Mauri? *Condor* 33:243–244.

———. 1947a. Charles Haskins Townsend. *Auk* 64:349–350.

———. 1947b. Vernon Orlando Bailey. *Auk* 64:502–503.

———. 1954. In Memoriam: Clinton Hart Merriam. *Auk* 71:130–136.

Parke, J. G. 1855. Report Explorations for that Portion of a Railroad Route near the Thirty-Second Parallel of North Latitude, Lying between Dona Ana, on the Rio Grande, and Pimas Villages, on the Gila. In *Reports of Explorations and Surveys . . . for a Railroad . . . to the Pacific Ocean.* 2, pt. 5:3–28. 33d Cong., 2d sess., S. Ex. Doc. 78.

Pattie, J. O. 1984. *The Personal Narrative of James O. Pattie*. Lincoln: Univ. Nebraska Press.

Peck, R. M. 1982. *A Celebration of Birds: The Life and Art of Louis Agassiz Fuertes*. New York: Walker.

Peterson, R. T. 1942. Bird Painting in America. *Audubon Mag.* 44:166–176.

———. 1965. In Memoriam: Ludlow Griscom. *Auk* 82:598–605.

Pfefferkorn, I. 1989. *Sonora*. Trans. T. H. Treutlein. Tucson: Univ. Arizona Press.

Phillips, A. M., III, D. A. House, and B. G. Phillips. 1989. Expedition to the San Francisco Peaks. *Plateau* 60, no. 2.

Phillips, A. R. 1939. The Type of *Empidonax Wrightii* Baird. *Auk* 56:311–312.

Phillips, A. R., J. Marshall, and G. Monson. 1964. *The Birds of Arizona*. Tucson: Univ. Ariz. Press.

Phillips, J. C. 1934. John Thayer. *Auk* 51:46–51.

Pope, J. 1855. Report of Exploration of a Route for the Pacific Railroad, near the Thirty-Second Parallel of North Latitude, from the Red River to the Rio Grande. In *Reports of Explorations and Surveys . . . for a Railroad . . . to the Pacific Ocean.* 2:1–50. 33d Cong., 2d sess., S. Ex. Doc. 78.

Price, W. W. 1888. Xantus's Becard *(Platypsaris Albiventris)* in the Huachuca Mountains, Southern Arizona. *Auk* 5:425.

———. 1895. The Nest and Eggs of the Olive Warbler *(Dendroica Olivacea). Auk* 12:17–19.

———. 1899. Some Winter Birds of the Lower Colorado Valley. *Bull. Cooper Ornith. Club* 1:89–93.

Purdie, H. A. 1879. The Golden-Cheeked Warbler and Black-Chinned Hummingbird in the United States. *Bull. Nutt. Ornith. Club* 4:60.

Ratti, J. T. 1979. Reproductive Separation and Isolating Mechanisms between Sympatric Dark- and Light-Phase Western Grebes. *Auk* 96:573–586.

Rea, A. M. 1983. *Once a River.* Tucson: Univ. Arizona Press.

Rehn, J.A.G. 1941. In Memoriam: Witmer Stone. *Auk* 58:298–313.

Rhoads, S. N. 1892. The Birds of Southeastern Texas and Southern Arizona Observed during May, June and July, 1891. *Proc. Acad. Nat. Sci. Phil.* 43:98–126.

Ridgway, R. 1876. Ornithology of Guadalupe Island, Based on Notes and Collections Made by Dr. Edward Palmer. *Bull. U.S. Geol. & Geog. Sur. Terr.* 2, bull. 2:183–195.

———. 1877. The Birds of Guadalupe Island Discussed with Reference to the Present Genesis of Species. *Bull. Nutt. Ornith. Club* 2:58–66.

———. 1880. Cataloque of Trochilidae in the Collection of the United States National Museum. *Proc. U.S. Nat. Mus.* 3:308–320.

———. 1884. Description of a New Species of Field-Sparrow from New Mexico. *Proc. U.S. Nat. Mus.* 7:259.

———. 1887a. The Coppery-Tailed Trogon . . . Breeding in Southern Arizona. *Auk* 4:161–162.

———. 1887b. The Imperial Woodpecker . . . in Northern Sonora. *Auk* 4:161.

———. 1887c. *A Manual of North American Birds.* Phil.: Lippincott.

———. 1901–1919. The Birds of North and Middle America. *U.S. Natl. Mus. Bull.* 50, pts. 1–8.

Ridgway, R., and H. Friedman. 1941–1950. The Birds of North and Middle America. *U.S. Natl. Mus. Bull.* 50, pts. 9–11.

Rivinus, E. F., and E. M. Youssef. 1992. *Spencer Baird of the Smithsonian.* Washington, D.C.: Smithsonian Inst. Press.

Roosevelt, T. 1924. Ranch Life and the Hunting Trail. In *The Works of Theodore Roosevelt.* Vol. 4. New York: Charles Scribner's Sons.

Rosenberg, K. V., R. D. Ohmart, W. C. Hunter, and B. W. Anderson. 1991. *Birds of the Lower Colorado Valley.* Tucson: Univ. Arizona Press.

Russell, S. M., and G. Monson. 1998. *The Birds of Sonora.* Tucson: Univ. Ariz. Press.

Sauer, G. C. 1982. *John Gould, The Bird Man.* Lawrence: Univ. Press Kansas.

Sclater, P. L. 1855. Note on the Sixteen Species of Texan Birds Named by Mr. Giraud of New York, in 1841. *Proc. Zool. Soc. London* 23:65–66.

———. 1857a. Notes on the Birds in the Museum of the Academy of Natural Sciences of Philadelphia, and Other Collections in the United States of America. *Proc. Zool. Soc. London* 25:1–8.

———. 1857b. On a Collection of Birds Made by Signor Matteo Botteri in the Vicinity of Orizaba in Southern Mexico. *Proc. Zool. Soc. London* 25:210–215.

———. 1857c. On a Collection of Birds Received by M. Sallé from Southern Mexico. *Proc. Zool. Soc. London* 25:226–230.

Scott, W.E.D. 1885. On the Breeding Habits of Some Arizona Birds. *Auk* 2:1–7, 159–165, 242–246, 321–326.

———. 1886a. On the Breeding Habits of Some Arizona Birds. *Auk* 3:81–86.

———. 1886b. On the Avi-Fauna of Pinal County, with Remarks on Some Birds of Pima and Gila Counties, Arizona. *Auk* 3:249–258, 383–389, 421–432.

———. 1887. On the Avi-Fauna of Pinal County, with Remarks on Some Birds of Pima and Gila Counties, Arizona. *Auk* 4:16–24, 196–205.

———. 1888. On the Avi-Fauna of Pinal County, with Remarks on Some Birds of Pima and Gila Counties, Arizona. *Auk* 5:29–36, 159–168.

Sennett, G. B. 1878. Notes on the Ornithology of the Lower Rio Grande of Texas, from Observations Made during the Season of 1877. *Bull. U.S. Geol. & Geog. Sur. Terr.* 4, bull. 1, art. 1:1–66.

———. 1880. Further Notes on the Ornithology of the Lower Rio Grande of Texas, from Observations Made during the Spring of 1878. F. V. Hayden. *Bull. U.S. Geol. & Geog. Sur. Terr.* 5, bull. 3, art. 21:371–440.

Serven, J. E. 1965. The Military Posts on Sonoita Creek. *The Smoke Signal* No. 12, pp. 26–33. Tucson: Tucson Corral of the Westerners.

Sharp, C. S. 1907. The Condor Fifty Years Ago. *Condor* 9:160–161.

Short, L. L., Jr. 1965. Hybridization in the Flickers *(Colaptes)* of North America. *Bull. Amer. Mus. Nat. Hist.* 129:307–428.

Shufeldt, R. W. 1887a. The Camera and Field Ornithology. *Auk* 4:168–169.

———. 1887b. Observations upon the Habits of *Micropus Melanoleucus,* with Critical Notes on Its Plumage and External Characters. *Ibis* 5th ser. 5:151–158.

———. 1914. Death of the Last of the Wild Pigeons. *Scientific Amer. Supp.* 78, no. 2024:253.

———. 1915. Anatomical and Other Notes on the Passenger Pigeon *(Ectopistes Migraturius)* Lately Living in the Cincinnati Zoological Gardens. *Auk* 32:28–41.

Sitgreaves, L. 1853. *Report of an Expedition down the Zuni and Colorado Rivers in 1851.* 32d Cong., 2d sess., S. Ex. Doc. 59.

Smith, A. P. 1907. The Thick-Billed Parrot in Arizona. *Condor* 9:104.

———. 1908. Destruction of Imperial Woodpeckers. *Condor* 10:91.

Southwestern Stockman. 1885. [Thick-Billed Parrots]. Vol. 2, No. 4, p. 4, col. 1, June 6, 1885; vol. 2, No. 19, p. 4, col. 1, Sept. 26, 1885. Willcox, Ariz.

Steinbeck, J. 1951. *The Log from the Sea of Cortez*. New York: Viking Press.

Stephens, F. 1878. Notes on a Few Birds Observed in New Mexico and Arizona in 1876. *Bull. Nutt. Ornith. Club* 3:92–94.

———. 1879. Nesting of *Buteo Zonocercus* in New Mexico. *Bull. Nutt. Ornith. Club* 4:189.

———. 1885. Notes on an Ornithological Trip in Arizona and Sonora. *Auk* 2:225–231.

———. 1899. Lassoing a California Vulture. *Bull. Cooper Ornith. Club* 1:88.

———. 1902. Owl Notes from Southern California. *Condor* 4:40.

———. 1903. Bird Notes from Eastern California and Western Arizona. *Condor* 5:75–78, 100–105.

———. 1918. Frank Stephens—An Autobiography. *Condor* 20:164–166.

Sterling, K. B. 1974. *Contributions to the History of American Ornithology*. New York: Arno Press.

———. 1977. *Last of the Naturalists: The Career of C. Hart Merriam*. New York: Arno Press.

Stevens, I. I. 1855. Memoranda in Reference to the Natural History Operations. In *Reports of Explorations and Surveys . . . for a Railroad . . . to the Pacific Ocean.* 1, pt. 1:9–10. 33d Cong., 2d sess., H. Ex. Doc. 91.

Stone, L. C. 1986. *Birds of the Pacific Slope*. San Francisco: Arion Press.

Stone, W. 1899a. Some Philadelphia Ornithological Collections and Collectors, 1784–1850. *Auk* 16:166–177.

———. 1899b. A Study of the Type Specimens of Birds in the Collection of the Academy of Natural Sciences of Philadelphia, with a Brief History of the Collection. *Proc. Acad. Nat. Sci. Phil.* 51:5–62.

———. 1901. John Cassin. *Cassinia* 5:1–7.

———. 1904. Samuel Washington Woodhouse. *Cassinia* 8:1–5.

———. 1907. Adolphus L. Heermann, M. D. *Cassinia* 11:1–6.

———. 1910. William Gambel, M. D. *Cassinia* 14:1–8.

———. 1916a. Henry Eeles Dresser. *Auk* 33:232.

———. 1916b. Philadelphia to the Coast in Early Days, and the Development of Western Ornithology Prior to 1850. *Condor* 18:3–14.

Streator, C. P. 1888. Notes on the Birds of the Santa Barbara Island. *Ornith. & Ool.* 13:52–54.

Streets, T. H. 1877. Contributions to the Natural History of the Hawaiian and Fanning Islands and Lower California, Made in Connection with the United States North Pacific Surveying Expedition, 1873–75. *U.S. Natl. Mus.* No. 7:9–33.

Stresemann, E. 1954. Ferdinand Deppe's Travels in Mexico, 1824–1829. *Condor* 56:86–92.

———. 1975. *Ornithology from Aristotle to the Present*. Cambridge, Mass.: Harvard University Press.

Sutton, G. M. 1979. *To A Young Bird Artist: Letters from Louis Agassiz Fuertes to George Miksch Sutton*. Norman: Univ. Oklahoma Press.

Swarth, H. S. 1904. Birds of the Huachuca Mountains, Arizona. *Pac. Coast Avifauna* No. 4.

———. 1905. Summer Birds of the Papago Indian Reservation and the Santa Rita Mountains, Arizona. *Condor* 7:22–28, 47–50, 77–81.

———. 1914. A Distributional List of the Birds of Arizona. *Pac. Coast Avifauna* No. 10.

———. 1926. James Hepburn, a Little Known Californian Ornithologist. *Condor* 28:249–253.

———. 1929. The Faunal Areas of Southern Arizona: A Study in Animal Distribution. *Proc. Calif. Acad. Sci.* 18:267–383.

———. 1934. In Memoriam: George Fean Morcom. *Condor* 36:16–24.

———. 1935. Review of D. M. Gorsuch's Life History of the Gambel Quail in Arizona. *Condor* 37:45–46.

Taber, W. 1958. In Memoriam: Charles Foster Batchelder. *Auk* 75:14–25.

Tanner, J. T. 1964. The Decline and Present Status of the Imperial Woodpecker of Mexico. *Auk* 81:74–81.

Taylor, H. R. 1887. A Trip to the Farralone Islands. *Ornith. & Ool.* 12:41–43.

Taylor, L. C. 1951. Prior Description of Two Mexican Birds by Andrew Jackson Grayson. *Condor* 53:194–197.

Thayer, J. E. 1906. Eggs and Nests of the Thick-Billed Parrot. *Auk* 23:223–224.

Thayer, J. E., and O. Bangs. 1907. Catalog of Birds Collected by W. W. Brown, Jr., in Middle Lower California. *Condor* 9:135–140.

———. 1908. The Present Status of the Ornis of Guadalupe Island. *Condor* 10:101–106.

Thompson, W. M. 1935. The Fighting Arizona Doctor. *Southwestern Medicine* 19:164–169, 181.

Thrapp, D. L. 1990. *Encyclopedia of Frontier Biography*. 3 vols. Spokane: Authur H. Clark.

Thwaites, R. G., ed. 1905. Account of an Expedition . . . to the Rocky Mountains . . . 1819, 1820 . . . from Notes of Major Long . . . by Edwin James. In *Early Western Travels*. Vols. 15 and 16. Cleveland: A. H. Clark.

Townsend, C. H. 1890. Birds from the Coasts of Western North America and Adjacent Islands, Collected in 1888–'89, with Descriptions of New Species. *Proc. U.S. Natl. Mus.* 13:131–142.

———. 1923. Birds Collected in Lower California. *Bull. Amer. Mus. Nat. Hist.* 8:1–26.

———. 1927. Oldtimes with the Birds: Autobiographical. *Condor* 19:224–232.

Townsend, J. K. 1837. Description of Twelve New Species of Birds, Chiefly from the Vicinity of the Columbia River. *J. Acad. Nat. Sci. Phil.* 7:187–193.

———. 1839a. List of the Birds Inhabiting the Region of the Rocky Mountains, the Territory of the Oregon, and the North West Coast of America. *J. Acad, Nat. Sci. Phil.* 8:151–158.

———. 1839b. *Narrative of a Journey Across the Rocky Mountains, to the Columbia River, and a Visit to the Sandwich Islands, Chili, &c.* Boston: Henry Perkins.

———. 1839c. Note on *Sylvia Tolmoei. J. Acad. Nat. Sci. Phil.* 8:159.

Uhler, F. M. 1951. In Memoriam: Albert Kendrick Fisher. *Auk* 68:210–213.

U.S. War Department. 1855–1860. *Reports of Explorations and Surveys, to Ascertain the Most Practicable Route for a Railroad from the Mississippi River to the Pacific Ocean.* 12 vols. Washington, D.C. Cited in text as *Pacific Railroad Reports.*

Van Tyne, J. 1929. Notes on Some Birds of the Chisos Mountains of Texas. *Auk* 46:204–206.

Vorhies, C. T. 1934. Arizona Records of the Thick-Billed Parrot. *Condor* 36:180–181.

Walsberg, G. E. 1993. History of the Condor. *Condor* 95:748–757.

Warren, E. R. 1936. Charles Edward Howard Aiken. *Condor* 38:234–238.

Weiss, H. B., and G. M. Ziegler. 1931. *Thomas Say: Early American Naturalist.* Springfield, Ill.: Charles C. Thomas.

Wetmore, A. 1931. Early Records of Birds in Arizona and New Mexico. *Condor* 33:35.

———. 1935. The Thick-Billed Parrot in Southern Arizona. *Condor* 37:18–21.

Wheelock, I. G. 1904. *Birds of California.* Chicago: A. C. McClurg.

Whipple, A. W. 1856. Itinerary. In *Reports of Explorations and Surveys . . . for a Railroad . . . to the Pacific Ocean.* 3, pt. 1. 33d Cong., 2d sess., H. Ex. Doc. 91.

Willard, F. C. 1923. The Buff-Breasted Flycatcher in the Huachucas. *Condor* 25:189–194.

Williams, T. 1987. *The Naming of Mount Graham.* Safford, Ariz: Graham Co. Hist. Soc.

Wilson, A. 1808–1814. *American Ornithology.* Phil.: Bradford and Inskeep.

Wilson, I. H., ed. and trans. 1970. *Noticias de Nutka.* Seattle: Univ. Wash. Press.

Winship, G. P. 1892–1893. The Coronado Expedition, 1540–1542. *U.S. Bureau Ethnology,* 14th Ann. Rpt. Washington, D.C.: GPO.

Woodhouse, S. W. 1852a. Descriptions of New Species of Birds of the Genera *Vireo,* Vieill., and *Zonotrichia,* Swains. *Proc. Acad. Nat. Sci. Phil.* 6:60–61.

———. 1852b. *[Sciurus Aberti]. Proc. Acad. Nat. Sci. Phil.* 6:220.

———. 1853. Report on the Natural History. In *Report of an Expedition down the Zuni and the Colorado Rivers in 1851 by Capt. L. Sitgreaves.* 32d Cong., 2d sess., S. Ex. Doc. 59:32–105.

Wright, A. H. 1912. Early Records of the Carolina Paroquet. *Auk* 29:313–363.

Wyllys, R. K. 1931. Padre Luis Velarde's Relación of Pimaría Alta, 1716. *New Mex. Hist. Rev.* 6, no. 2:111.

Wynne, O. E. 1969. *Biographical Key—Names of Birds of the World—To Authors and Those Commemorated.* Courtwood, Sandleheath, Fordingbridge, Hants: Col. O. E. Wynne.

Xántus, J. 1858. Descriptions of Two New Species of Birds from the Vicinity of Fort Tejon, California. *Proc. Acad. Nat. Sci. Phil.* 10:117.

———. 1860a. Catalogue of Birds Collected in the Vicinity of Fort Tejon, California, with a Description of a New Species of Syrnium. *Proc. Acad. Nat. Sci. Phil.* 11:189–193.

———. 1860b. Descriptions of Supposed New Species of Birds from Cape St. Lucas, Lower California. *Proc. Acad. Nat. Sci. Phil.* 11:297–299.

Zwinger, A. H. 1986a. John Xántus: The Fort Tejon Letters 1857–1859. Tucson: Univ. Ariz. Press.

———. 1986b. *The Letters of John Xántus from Cabo San Lucas.* Los Angeles: Dawson's Bookshop.

INDEX

ABOUT THE AUTHOR

DAN L. FISCHER was born on June 17, 1932. His close association with the borderland region began after he arrived as a young boy with his parents in Yuma, Arizona, in 1937. Although his professional career was that of an engineer in industry, his preoccupation, or second vocation, has clearly been related to natural history subjects with a strong emphasis on birds. For over fifty years, Fischer has traveled the borderlands, pursuing and photographing birds, while retracing the journeys of many early explorers and naturalists. He expanded these interests to become a serious student of the historical aspects of ornithology during the last several years. Following a suggestion to write about early naturalists, Fischer reviewed several hundred journal entries, manuscripts, and publications on the topic, and a book on this long-neglected subject finally took form. He has been gathering material with the possibility of doing a similar book on another region. Fischer's photographs have illustrated many magazines and books. He is a member of several bird organizations, including the Point Reyes Bird Observatory, Southeast Arizona Bird Observatory, National Audubon Society, American Birding Association, HawkWatch International, Texas Ornithological Society, and since 1958, the Cooper Ornithological Society.